-2-64: Weather today very pleasant — temp. about 32°, a few flakes of snow. Wind almost dead calm all day, but around 5:00 PM picks up from the south. Clouds high scattered cirrus, sometimes lower + thicker. Ice — almost no ice in sight except those calves grounded near the village and north of it.

Most of the men, as mentioned before, have gone to Barrow for about 2 weeks, the other men are either working around the village or are inside. I am beginning to realize that the people are definitely reserved toward me, and that it will be some time before I am [illegible] as I hope to be able to. It [illegible] having some role to play, [illegible] really no role until the [illegible]

They say that the ice [illegible] about 2 weeks ago, and [illegible] for hunting will not form until around November or December. ~~All of the men~~

All of the men appear to have heard about my qajaq, and if we talk about it there is nothing I can tell them that they haven't heard previously from Alva or Alfred. The store owner, Raymond Aguvlak, is said to be a good qajaq man, and appears especially interested in it.

Today I looked at a qajaq on a stand in the village. I was astonished at the size, only about 10 feet ~~or~~ 12 feet in length, width about 22-24 inches. Will measure and sketch later, when I become friendly with the owner. General design is this:

Has many stringers + bulkheads inside, white skin covering.

PRAISE FOR *RAVEN'S WITNESS*

"Hank Lentfer's *Raven's Witness* is an exceptional biography of Richard Nelson, and a gift to us. Reading it we come to understand, through Nelson's life, how the immediacy of the senses, flowing across the earth—touch, taste, scent, sight, sound—transcend whatever other feeble boundaries we construct. Nelson emerges here as a most passionate participant in these senses, and in that immediacy, endures."

—Rick Bass, author, *Winter: Notes From Montana*

"An anthropological saga, a literary gem about the legendary 'Nels,' one of the most extraordinary naturalists of our time."

—Paul Hawken, author, *Drawdown*

"In *Raven's Witness*, Hank Lentfer pays homage to one of the most talented naturalists and field anthropologists of the last century, Richard Nelson. The book is a double treat: an intimate portrait of the exuberance and joy Nelson brought to his life, as well as an eloquent and distinctive story of the terrains and cultures that both subject and author explored."

—Gary Paul Nabhan, author, *Food from the Radical Center*

"I savored every page of this magical, beautifully woven book—the story of one man's apprenticeship to Alaska's northern wisdom and wild joy. The final pages brought me to laughter and tears, inspired by Richard Nelson's humility and respect. His great heart forever open—listening."

—Kim Heacox, author, *The Only Kayak*

"Even more than a brilliant biography, *Raven's Witness* creates an entirely new genre—a beautifully told story of the ethos of gratitude and joy that is born when an extraordinary man immerses himself in the culture of sea-ice people and the wisdom of a wild island."

—Kathleen Dean Moore, author, *Great Tide Rising*

"*Raven's Witness* is the Alaska book I've waited lifelong to read. Hank Lentfer's gorgeous account of the evolution of Richard K. Nelson's life and thought is a godsend. Half a century ago Nelson realized how wrong it was to send Native Alaskans *our* teachers when we should be begging theirs to teach us, and Lentfer describes the ways he came to live that realization. Nelson, in the end, was no kind of professor, prophet, or scientific professional: he was a joystruck participant in a sentient and seeing world in which 'no one is ever alone, unseen, or unheard' and 'gratitude kindles the very heat of life.' May Earth and the Unseen bless every woman, man, and child drawn to such a path."

—David James Duncan, author, *The River Why*

RAVEN'S WITNESS

THE ALASKA LIFE OF RICHARD K. NELSON

HANK LENTFER

FOREWORD BY BARRY LOPEZ

MOUNTAINEERS BOOKS

MOUNTAINEERS BOOKS is dedicated to the exploration, preservation, and enjoyment of outdoor and wilderness areas.

1001 SW Klickitat Way, Suite 201, Seattle, WA 98134
800-553-4453, www.mountaineersbooks.org

Printed in Canada
Distributed in the United Kingdom by Cordee
23 22 21 20   1 2 3 4 5

Copyeditor: Elizabeth Johnson
Design and layout: Jen Grable
All photographs by the author unless credited otherwise
Cover and interior feather image © istock/StudioDoros
Cover and frontispiece photograph: *Richard on young ice, with open water in the background. Rifle over his shoulder, he holds an unaaq used to test ice thickness. The umiaqaluraq is used to retrieve seals.* (Nelson archive photo)
Back cover photograph: *Iñupiaq hunters pull an umiak over thin ice after retrieving a seal.* (Nelson archive photo)
Back flap photograph: *Richard and Hank take a break after a morning recording birds in Glacier Bay.* (Photo by Liz McKenzie)

Library of Congress Cataloging-in-Publication data is on file for this title at https://lccn.loc.gov/2020004651

Mountaineers Books titles may be purchased for corporate, educational, or other promotional sales, and our authors are available for a wide range of events. For information on special discounts or booking an author, contact our customer service at 800-553-4453 or mbooks@mountaineersbooks.org.

Printed on 100% recycled paper and FSC-certified materials

ISBN (hardcover): 978-1-68051-307-3
ISBN (ebook): 978-1-68051-308-0

An independent nonprofit publisher since 1960

"We're here for a little window. And to use that time to catch and share shards of light and laughter and grace seems to me the great story."

—Brian Doyle.

CONTENTS

FOREWORD

It seems appropriate for me to reflect first on the undistinguished chair I'm sitting in as I try to put together a few words to introduce you to this biography of Richard Nelson. I bought the chair long ago in a secondhand store in Springfield, Oregon. I've had to repair it occasionally, to ensure its sturdiness. Two worn-out seat cushions, one atop the other, make it easier to occupy for hours at a time. Two newel posts brace a tapered backrest of wooden spindles. The caps of the newel posts gleam from the rub of human hands over the decades.

I've written seventeen books sitting in this chair, and I hope to complete a couple more in the years ahead. In the early 1980s, because I sensed that resting my back against a pair of cured blacktail deer hides from Richard's hunts would put me in a more respectful frame of mind when I wrote, and that they might induce in me the proper perspectives about life, I wrote him and asked for his help. Would he honor our friendship by sending me a couple of hides? These were from blacktail deer he'd hunted in the woods near his home. (The way he understood these fatal encounters, as a subsistence hunter trained by indigenous Koyukon people, he'd been "given" these deer.) In my experience no other non-Native hunter's ethical approach to this archetypal encounter with a wild animal was as honorable as Richard's. He hunted to feed his family, imitating the way his Iñupiaq, Koyukon, and Gwich'in teachers had taught him to, through the example of their own behavior in engagements with wild animals—honoring, humble, grateful. I felt the hides might care for me as I stumbled my way through life, in the same way that our friendship with each other would take care of both of us in the years ahead.

So, this morning, I am sitting in front of my old typewriter with my back against those soft hides, and I want to say a few things about this book in which Hank Lentfer introduces us to this extraordinary man, Richard K. Nelson, most of whose friends called him simply Nels. I think of him as remarkable among Western anthropologists who have apprenticed themselves to traditional people, because, as a young man, he took on—with fierce dedication in discouraging circumstances—a journey few people have ever had the opportunity to experience, and he pursued that journey with great attentiveness and care for more than fifty years. He listened to his teachers, immersed himself in their landscapes as a naturalist, and became, without intending to, a great teacher himself. What Nels modeled was a way of knowing the world, an epistemology different from the one most of us have unconsciously accepted and sworn ourselves to. It led him to reconsider what in his own life was to be valued over the manic accumulation of material wealth, or progress for the sake of progress, or the exceptional status of humans among all other beings.

You'll read here in Hank's thoroughly researched chapters about a man who dedicated himself to an unending personal task—self-education—and who became, because of that, to use Lentfer's unusual but apt term, a kind of monk. He didn't pursue life alone in a cave, staying out of touch with others and with the physical world, but became instead a carrier of wisdom from traditions he had known nothing about as a young university student. This is to say the wisdom of marginalized and vanishing hunter-gatherer cultures in remote parts of North America. He came to understand, through decades of intimacy, applied study, and life practice, a frame of mind he was once innocent of, one based much more firmly on enduring human values like thinking morally and ethically about wild animals. As he wrote about this perspective in books and articles, and as he began to speak publicly about other ways of knowing, he developed the aura of a person shaped by something on the far side of pedestrian reality. A monk, then.

A key to understanding these pages is to recognize how often illumination came to Richard in the company of other people, to see the social dimension of the wisdom he offered to share with us. Over time, Nels slowly became someone more intent on listening than in advancing himself as a speaker. Iñupiaq people in the 1960s saw that this outlander from Wisconsin was willing to listen attentively and to work hard in order to acquire knowledge and field skills. By the time he arrived among the Koyukon in Huslia and Hughes in the '70s, he had learned to present himself as quintessentially a listener, not a cultural anthropologist. As a result, the breadth of his learning widened and deepened. By the time he was camping in Glacier Bay with Hank Lentfer in 2016, he was listening to the world around him so closely he could differentiate between the seemingly identical songs of two fox sparrows calling from the same thicket. He could tell from the anxious behavior of a gray whale in a protected bay on Kruzof Island that orcas were swimming nearby, unseen. He understood that his professional position in life had moved beyond anthropology, good as he was at that, to a level of attentiveness to the nonhuman world that one of his Koyukon teachers, Lavine Williams, alluded to when he informed Nels one day in Hughes that "every animal knows way more than you do."

The truth of Lavine's remark was both metaphorical and factual. To be patient, to pay attention to the world that is not you, is the first step in the neophyte's discovery of the larger world outside the self, the landscape in which wisdom itself abides. Other residents of the world, Lavine was telling Nels, know more than we do about how to survive whatever is coming. Their investment is not in progress but in stability. Nels's life became the passing on of this deceptively simple message about human survival. It was this message that he lived.

—Barry Lopez
Finn Rock, Oregon
November 2019

PROLOGUE: SOLID GROUND

My earliest memories are frozen on the Chukchi Sea. A bowhead whale had been killed and my family, like everyone else in Utqiagvik, traveled to the ice edge to lend a hand. I don't remember breakfast, getting dressed, or climbing onto the sled behind my father's snow machine. But I recall a wrist-thick length of twisted rope, coal black against the snow. The line ran through wooden pulleys the size of hubcaps. One end was anchored in a hole drilled through the ice. The other end disappeared into smooth, dark water.

Everyone—children, grandparents, men, women—assembled along the rope. Even I, a four-year-old white kid, more hindrance than help, was allowed to join the line.

"One . . . two . . .three . . . heave!"

The line tightened, the pulleys squeaked. Heels dug in, forearms burned, people leaned against the weight. Inch by inch, we backed up, and a whale emerged, tailfirst, from the sea.

When the rope went slack, we all crowded around. Atop the ice lay the animal's full bulk—as long as a bus, taller than a man. Blood seeped from beneath the whale's head. Its massive pink tongue lolled between the comblike rows of baleen. Images in our family photo album sharpen the hazy edges of my memory. Pictures of men standing on the whale's back, cutting wide slabs of blubber with long-handled knives. Women serving steaming platters of *maktak*, strips of black skin and white fat. Smiles and laughter all around; good fortune, like a warm wind, touched everyone.

As a kid, I was ignorant of the deep currents and shifting winds capable of buckling and fracturing the sea ice beneath our feet. I simply trusted the adults to keep me safe. I'll never know who noticed the crack that threatened to set us all adrift. But I'll always remember celebration swamped by fear.

It was a frantic, bumpy ride back to town. Along with my mother and two sisters, I clung to the bucking sled as Dad raced along the rough trail chopped through jumbled ice. I remember two men helping each sled navigate the crest of a steep pressure ridge. When I was pitched from my father's sled, one of those men tossed me in with the next family. Someone caught me, held on, delivered me to solid ground.

It's been nearly fifty years since I last experienced a *naluktaq*, the multiday party that swept through Utqiagvik following each whale harvest. I can still taste the oily bounce of maktak and recall the rush of the blanket toss—people thrown impossibly high, arms spinning, backlit by clouds. I can still hear the snap of sticks on skin drums and feel the rhythmic stomp of dancers' feet.

We all, I believe, hunger for such close-knit community. We yearn to pull together with neighbors and celebrate our collective success.

So why do we find ourselves living in such rancorous times? How did stories of unity get buried by the din of voices tearing us apart? When did caring for our country become a partisan issue? And by country, I don't mean a flag, song, or pledge but our actual home ground—the soil, rivers, forest, tundra, air, and climate that make life possible.

In the introduction to his book *Tribe: On Homecoming and Belonging*, Sebastian Junger writes: "Humans don't mind hardship, in fact they thrive on it; what they mind is not feeling necessary." As a boy tugging on that whale, I felt, for a moment, a solidarity that stretches back through our evolutionary past. "A wealthy person," writes Junger, "who has never had to rely on help and resources from his community is living a privileged life that falls way outside of more than a million years of human experience."

Our economy feeds the notion that my well-being is independent of yours. In the belief that no one else will feed you, it's prudent to hoard against starvation. If there's not enough to go around, it makes sense to build walls, to take another's wealth and add it to your own. But no amount of money can bury the truth of our interdependence. Lost in the scramble to build barriers and bank accounts is the richness found in working together.

In her book *A Paradise Built in Hell*, Rebecca Solnit explores how disasters, both human-made and natural, shatter the patterns that isolate and divide. In towns flattened by winds, in cities leveled by bombs, people emerge from the rubble and become their brother's keeper. In helping each other survive, she writes, they find a "purposefulness and connectedness that brings joy even amid death, chaos, fear, and loss."

When the ice cracked, my Iñupiaq neighbors sought solid ground. But how do we respond to the widening split between economic desire and ecological reality? Where is solid ground when coal plants in Kentucky deepen droughts in Kenya? What's the response when local depletions merge into global extinctions?

We cannot bridge the growing gulf between economics and ecology until we see personal well-being as inseparable from the planet's health. We must pull together, expanding the circle of caring beyond immediate family and neighbors.

"Empathy," the fourteenth Dalai Lama has said, "is the basis of human coexistence. It is my belief that human development relies on cooperation, not competition."

So how do we widen empathy to include strangers? How do we swell our affinity for community to include other creatures?

Stories.

We need better stories.

The stories of separation that got us into this mess are not the stories that will get us out. We need stories that illuminate the truth of connections—to each other and to this precious blue planet, our only home. We need to reject narratives of division. We need storytellers who blur boundaries, expand empathy, and stretch our capacity for caring.

Richard K. Nelson is one such storyteller.

And, like any good storyteller, he had to first learn to listen.

Richard was a keen cultural anthropologist and an astute naturalist. What began as a single winter on the windswept tundra opened into a lifelong journey of listening. For five decades, he mushed along frozen rivers, wintered in log cabins, cooked meals over seaside fires. He listened to Iñupiaq hunters puzzle over the intricacies of shifting sea ice and migrating caribou. He lived with Koyukon Indians moving through a forest of eyes. He reveled in the whisper of wind through feathers, the scratch of talons on bark, the hiss of rain through trees. And along the way, he took notes. Thousands of pages of notes. By candlelight or gas lantern, he wrote each evening until his hand cramped, night after night, year after year.

Those notes grew into books, which now line my shelf. His original spiral-bound journals fill boxes jumbled around my desk. The thirteen thousand plus pages chronicle a transformative journey. A journey through loneliness to a deep sense of home.

Throughout this book are entries from many of Richard's journals, which show this transformation, like this passage from his Sitka-based Rain Journal:

> *There is nothing in me that is not of earth, no split instant of separateness, no moment of existence in which a single particle sets me apart from earth. We have forgotten the most basic and elemental truth of all.*
>
> *Until this truth becomes intuitive again in the Western world we jeopardize our own existence. Others have not forgotten, and we must go back to them seeking after lost knowledge.*

In researching this book, I interviewed dozens of people who have known Richard over the years. Author Barry Lopez was one of them. Before having dinner with Lopez, I read his *The Rediscovery of North America*. A profound book, small only in its size, it opens in the early hours of October 12, 1492, when Juan Rodriguez Bermeo, aboard the *Pinta*, spots a dark curve of land rising from the moonlit sea. Bermeo's discovery soon has his sleepy shipmates scurrying on deck as their captain, Cristoforo Colombo, gives orders to reef the sails and drift until dawn.

"What wealth," Lopez asks, "did Bermeo cry out in anticipation of? Let us be kind, as we hope historians will be kind to us, and say that for him and for many others on those three small ships perhaps the hummingbird would have been enough. The hummingbird, fresh food after five weeks at sea, and the astonishing lives of the Arawak."

Sadly, we now know those sailors would not be satisfied by natural marvels and new cultures. "This incursion . . . into the 'New World,'" Lopez writes, "quickly became a ruthless, angry search for wealth. It set a tone for the Americas. The quest for personal possessions was to be, from the outset, a series of raids, irresponsible and criminal, a spree, in which an end to it—the slaves, the timber, the pearls, the fur, the precious ores, and, later, arable land, coal, oil, and iron ore—was never visible, in which an end to it had no meaning.

"The assumption," Lopez continues, "that one is *due* wealth in North America, reverberates in the journals of people on the Oregon Trail, in the public speeches of nineteenth-century industrialists, and in twentieth-century politics." What would have happened, Lopez wonders, if Bermeo and his shipmates had come ashore asking, instead: Who are these people? What is this land?

Stories are born from questions. And the questions we ask matter. They shape how we see the world. And the world is shaped by our vision.

Five hundred years after those ships arrived, a kid from Wisconsin harnessed up a ragtag team of sled dogs, joined a group of Iñupiaq hunters, and rediscovered North America. *Who are these people? What is this land?* Where others had scrounged for gold—and would soon drill for oil—Richard K. Nelson mushed across the solid, shifting sea, and found his way home.

PART I

Niglik

CENTER OF GRAVITY

Under the plane's right wing, bright islands of ice floated on a steel-gray sea. Under the left wing, shining puddles and ponds spilled across the flat brown tundra. Shoulder to shoulder with the pilot, a young man in the passenger seat swung his gaze from side to side, mesmerized with his first look at this strange world. The pilot flew with two fingers on the yoke, keeping the single-engine Cessna a mere hundred feet above the beach.

The pilot leaned over and shouted above the engine's roar, "First time up this way?"

"Yeah," yelled the young man. "This is fantastic!"

"Ever seen a caribou?"

"Not yet, but I sure want to."

The plane banked inland and lost altitude. Vivid patterns of water and land flashed by, now only thirty feet below.

"Just ahead," barked the pilot.

Soon they overtook a small herd. The frightened animals ran and swerved. The pilot buzzed a dozen groups, scattering each before lifting the plane away.

"There. Now you've seen 'em."

"Holy smokes!"

The young man was twenty-two years old, still a kid really. Forehead pressed to the window, he studied the landscape. No trees, no mountains, no buildings, no roads. Just plates of ice, silent and still. Rivers coiled like snakes.

A half hour later, a cluster of low buildings slid into view. The plane buzzed the village, and the young man glimpsed his new home, snugged near the coast. The pilot made a sharp seaward turn, right wing tip toward the water, then leveled the plane just above the beach. It hit hard and popped back into the air, bounced twice more, and quickly slowed in the soft black sand.

The pilot killed the engine and grinned. The young man's ears rang in the sudden silence.

"Last stop," said the pilot. "Everybody out." He unbuckled, jumped from the plane, and began unloading the jumble of gear. He handed each parcel to the young man.

The young man's features were classically Nordic: cropped blond hair and soft blue eyes, bright, even teeth and a strong, clean jaw. He moved with a liquid grace, his slim frame tuned with strength and balance as he stacked the boxes and duffels in a growing heap.

Last to emerge from the plane was a rifle in a cloth case, and two long, awkward bundles containing a folding kayak.

"Good luck." The pilot hopped into the cockpit. "Oh, and shield your eyes when I take off."

The engine roared, engulfing the man in wind, sand, and noise. The plane rumbled down the beach, eased aloft, and slipped into the wide gray sky.

A stiff breeze off the Arctic Ocean slapped at the young man's blue jeans and coat. A loose box lid rattled. Small waves lapped the black sand.

Inland, a steep bluff rose fifteen feet above the shore. Atop the bluff, a growing line of people stood motionless, staring down at the stranger. The young man gazed back.

It was the last day of August 1964 in the village of Wainwright, home to 290 Iñupiaq Eskimos and 540 sled dogs. The young man was Richard K. Nelson, an anthropology student at the University of Wisconsin in Madison. His parents and older brother knew him as Richard. His high school chums and college buddies referred to him by Nels or the common nickname Dick. The Eskimos atop the bluff would soon give him a new name.

Richard had no way of knowing the dark days of the coming winter would hold the most lonesome and lovely moments of his life; that the language and lessons, teasing and tenderness, of the men on the bluff would form a gravity around which the rest of his life would bend.

He hefted a box of food and trudged toward the line of people. Somewhere a husky yipped, then broke into a howl. Another dog answered.

Hundreds of huskies joined in, their combined voices loud against the wind.

A dozen children clad in fur parkas clustered where the path crowned onto the flat tundra. Most wore sealskin mukluks. A few had rubber boots.

"What's yer name?" asked the boldest of the boys.

"Dick."

The kids glanced at each other and giggled.

"Dick-Jane-Sally," exclaimed the bold boy.

"Dick-Jane-Sally, Dick-Jane-Sally," echoed the others, spiraling themselves into a storm of laughter.

Richard flinched, afraid they were making fun of him. Then he realized the joke: to the children, it was as if the main character from the ubiquitous Dick and Jane reading primers had just walked into town.

"What's *your* name?" Richard asked the bold boy.

"Norman. Norman Matoomealook."

"Do you know where the Arctic Research Lab house is?"

"Come on. I'll show you." Norman led the way to a pale-yellow building a few hundred feet from the bluff. Richard stepped through the door into a tight entryway, with just enough room to turn around. Coat hooks poked from the wall. A roll of toilet paper hung from a nail beside a plastic bucket.

Richard pushed through the next door and set the box of food on the only table. He stood a moment and let his eyes adjust to the dim light filtering through three small windows. Along the seaward end of the twelve-by-twenty-foot room stood a counter and a small four-burner stove. No sink. Two chairs pulled tight to the table. Metal bunks at either side of the door. A single gas lantern hanging from the ceiling.

When he left for more gear, he approached a small group of adults lingering at the bluff. The women replied to his greeting with a simple "Hi." The men responded with a slight nod of the head. Richard made over a dozen trips, lugging gear from the beach. No one helped.

In the following days, temperatures hovered just above freezing, skies high gray, wind stiff from the northeast. He spent hours wandering the web of muddy paths connecting the small single-story homes scattered about the village. Few had more than one tiny window per wall. All had simple gable roofs and entryways framed out from the doors. Unlike Madison neighborhoods, where all the houses faced the street, the Wainwright homes pointed in all directions, some houses tight together, others alone on the edge of town.

Each yard had a half dozen to twenty dogs tethered on short chains. Head-high wooden racks, cobbled together with driftwood and odd boards, kept an assortment of skin boats, seal hides, and caribou legs out of reach of loose animals. Dogsleds, awaiting the first snow, leaned against buildings or rested on racks. Odd boards, walrus skulls, caribou antlers, garbage sacks, and other debris lay scattered between the homes. People glanced from doorways and yards but kept their distance from the wandering stranger.

Richard was grateful when Norman stopped by to invite him ptarmigan hunting. With a .22 rifle slung over his shoulder, Richard's new friend led the way out of town, scampering over uneven tussocks. Aside from knee-high willows bristling from low gullies, the vegetation was no more than ankle-deep. Nothing to break the wind. Richard raised his collar and trudged on, happy for the company of this genial boy so comfortable on the stark tundra.

They stopped where the Kuk River cut inland, flooding the tundra with a wide lagoon. They watched ducks, geese, and loons swimming just beyond the reach of Norman's gun. When a gull flew over, the boy drew it near with a clucking call. With a single shot, the bird pinwheeled down.

"Dog food," Norman explained, holding the gull by its feet.

Within a few years, the howl of huskies would be replaced by the whine of snow machines. But in 1964, all overland travel was on foot or on the runners of a dogsled. In the coming months, Richard would learn firsthand anything that swam, walked, or flew was potential food to keep the huskies nourished and pulling.

Back in the village, Norman invited him home for dinner.

Richard had to bend almost in half to ease through the two-by-three-foot doorway. Inside, his hair brushed the low ceiling of the cluttered one-room house. A counter of rough boards stretched along a wall beneath a small dust-caked window. A caribou-skin-covered bed stood in one corner. In the room's center was a wooden table and two well-used chairs.

Norman's radiant, round-faced mother, Lena Mae, smiled in greeting, her high cheekbones pushing her eyes into narrow slits.

"*Azaa*, you're so tall," she said.

Richard chuckled. As the smallest kid in his high school, he'd been relentlessly teased about his size. In college, he grew to a respectable five foot ten but still thought of himself as the class runt.

"Sit, sit," Lena Mae said.

Richard joined Norman at the rough table. Lena Mae served two chunks of frozen raw caribou and a bowl of stinking amber liquid. Following Norman's lead, Richard picked up a knife and shaved thin strips from the frozen meat. But when Norman dunked the meat into the bowl of putrid fluid, Richard hesitated.

Lena Mae laughed. "Seal oil. Go ahead. It's good," she said.

Richard dipped some raw caribou, held his breath, and brought it to his mouth.

Later, in a letter to his parents, he described the penetrating odor of seal oil.

> *It's intense, but I've gotten used to it. Kind of tasteless, actually. Like dipping your food in vegetable oil—except for the fermented smell. It was a little bit frightening but I went ahead. It wasn't bad.*

Richard was lucky to have been invited into someone's home so quickly. The people of Wainwright had, in recent decades, grown accustomed to white people coming there to teach and preach. But outside the single school or one of the two churches, a white man's presence was suspect. In another letter, he confided:

> *Dear Ma and Pa,*
>
> *I have been having considerable difficulty getting friendly with the adults, who seem pretty unsure about me, and perhaps suspect me of being some kind of Fish and Game Dept. spy, secret missionary or something.*

Wainwright had one village store: a single room with rough plank flooring and metal shelves half filled with pilot bread, candy, tea, and canned goods. Raymond Aguvluk, the store manager, was the first to flat-out ask what Richard was up to.

Richard explained the US Air Force wanted a sea-ice survival manual for pilots flying missions over the Arctic and the Antarctic. His job was to interview locals, travel on the ice, and send monthly reports to a cold-weather survival research facility in Fairbanks.

Raymond listened, nodded, and, with hands clasped behind his head, leaned back and said, "How come they sent you? You don't know nothing."

BUFFALO AND BULLFROGS

Richard was born on the first of December 1941, in Madison, Wisconsin. Walt Disney had just released the animated film *Dumbo*. General Mills introduced the breakfast cereal Cheerios. The average price of a new home was $4,150. Gas was nineteen cents per gallon.

Richard's father, Robert Nelson, managed the office of unemployment compensation for the state of Wisconsin. Florence Nelson was a homemaker. From their house on Outlook Street, Flo and Bob looked across the calm waters of Lake Monona. Wednesday evenings, they bowled in the local city league. Sundays, after the Lutheran service, they tuned their radio to the Green Bay Packers game. In the Nelson house, Jell-O with mini marshmallows was the norm, ketchup an exotic spice. Salad was iceberg lettuce sprinkled with sugar. This was Norman Rockwell's America. This was *Leave It to Beaver* brought to life.

Flo was just home from the hospital with her newborn when the radio described columns of smoke rising from the ruin of Pearl Harbor. The United States declared war on Japan on the eighth of December. Richard had been alive for a week.

Bob was a gentle father and generous neighbor. For years, he came home from work, ate dinner with his family, and then put in an evening shift building batteries to help the war effort. Bob suffered from chronic asthma, and the added hours at the factory stressed his fragile health. His movements were slow, each step judiciously spent.

Alongside Bob's cautious pace, Flo whirled about with an abundance of nervous energy. Richard was just three months old when his mother moved the family out of the house on Outlook Street into a cheaply priced multistory home across Lake Monona, on Few Street. Throughout the war and for years afterward, Flo bought, remodeled, and prettied up house after house at a pace that moved the family through sixteen homes in the first sixteen years of Richard's life.

The home that Richard most remembers, the one he wishes had never been sold, was twenty miles east of Madison, in the farming community of Marshall. While Bob commuted into the city, Flo focused her tornado of energy into running a classic small-town store. Nelson's General Store sold it all: coveralls and motor oil, shovels and bread. The family lived above the store for Richard's fourth- and fifth-grade years.

Richard did well at Marshall Elementary. His report cards filled with As and Bs, marred only by a string of Ds in math. Outside of school, he rode along with the neighboring farmer, Jim Herman, as he mowed hay. He also helped Farmer Jim in the barn, hauling feed for the cows and shuttling pails of frothy milk. But more than in the small school and the expansive barn, Richard flourished in the open air.

Gripping a fishing pole across the handlebars of his bike, he pedaled off on summer mornings to meet up with friends from surrounding farms. They caught bluegills and crappies in the ponds and lakes. They shucked clothes to skinny-dip in the afternoon heat and snuck up on frogs, pounced on snakes. It was a Huck Finn life: always another tree to climb, always a distant patch of woods to explore.

The Maunesha River flowed lazily behind the house. Richard would have been happy to spend the rest of his childhood fishing those slow waters. But Bob, his asthma growing worse, was tiring of the commute from Marshall. And Flo, itching for the next challenge, sold Nelson's General Store and bought another home near Lake Monona. As his folks shuttled carloads of possessions into town, Richard stayed behind. He spent those final hours on the riverbank, casting into the current. When the time came, he slumped into the back seat and cried all the way to Madison.

The curiosity that had thrived in Marshall's open fields shriveled in the cramped halls of East Junior High. A line of Ds in English, math, and speech riddled Richard's seventh-grade report card. This dismal performance was in stark contrast to the As and Bs earned by his older brother. Yet Bob and Flo never pressured their younger son to change his ways.

"We know you're smart," they said. "We know you could do better if you wanted to."

When he wasn't in school, Richard prowled Madison's sloughs and lakeshores in search of turtles and snakes. By the time he enrolled in Monona Grove High, his outside interests had swelled into an obsession. As his classmates did homework, he devoted every scrap of daylight to pursuing creatures wiggling along the edges of ponds and the bottoms of ditches. Anything that scuttled, skittered, or slithered was to be chased, captured, and studied. His parents reluctantly indulged their son's passion, allowing Richard to fill his room with crates, boxes, and tanks to hold his revolving cadre of pets.

Naturalist was not part of Richard's boyhood vocabulary. His mom was terrified of snakes. His dad preferred the comfort of the couch over the tangle of the woods. His older brother, Dave, was more interested in radios than reptiles. But Richard was caught in a current of curiosity he could not escape. The more creatures he found, the harder he looked.

Flo eventually parked in the driveway and relinquished the garage to her son's growing array of aquariums and cages. On weekends, Bob drove out of town and sat in the car while his son prowled riverbanks and forest edges for newts and salamanders. Richard appreciated the support, but it would be years before he lived with people who shared his growing interest in the intricacies of the natural world.

Out of Richard's flock of friends, Gerald Burgette was the only one who shared his fascination with reptiles. The two made weekly trips to the pet store to buy mice for the hungry snakes. They captured crickets for the frogs, snuck lettuce from the fridge for the turtles, and freshened the bowls of water for the salamanders.

When Gerry's family up and moved to Tucson, the boys' obsession spanned the distance. Gerry was soon capturing desert snakes and dryland lizards and mailing them back to his friend in the Midwest. Gerry packed each captive in the hollow center of a cholla cactus stick and sealed the ends with newspaper. Richard freed the animals in the bathtub, never sure what would come skittering out.

As his collection of critters grew, his interest in school waned. His ninth-grade English teacher, Mrs. Stelter, wrote on the back of his report card, "Richard needs to exercise a little self-control in class."

After barely passing him in tenth-grade geometry, Miss Frost wrote, "Richard's grades are very low. In tests he has had 7, 20, 80, 7, 41, 100, 0, 0." The perfect score had come the day Richard was sitting behind his math-whiz buddy Pete Baime. Miss Frost separated the boys, and Richard flunked the next two tests. "In the last exam," wrote Miss Frost, "he didn't do a single problem. He seems to be interested only in attracting attention instead of paying attention to the explanations I give in class."

Richard wasn't rebelling. He didn't refuse to study. He simply wasn't curious about solving equations or diagramming sentences. He felt bad handing those report cards to his folks. But he also felt powerless to generate interest where none existed.

Bob and Flo could have hired a tutor or sent Richard to summer school. They could have made him sit indoors and study until his grades improved. Instead, they sent their son to visit his lizard-loving friend in Arizona. Richard climbed aboard a Super Constellation, a four-engine plane, for the first flight of his life. After a week of desert adventures, he said goodbye to Gerry in the Tucson train station and settled into a window seat for the long ride home.

As the train climbed into the Rockies, he watched the occasional cactus give way to scattered forest. Nothing in glacially scoured Wisconsin had prepared him for the spectacle of Colorado's jagged peaks. He was in high spirits as the train rattled down the foothills and set a straight course across the plains.

For years, Richard had combed library shelves for books about critters and plants. He'd repeatedly returned to a book describing the tall grasses swaying across the Great Plains. He wandered through those pages into the convoluted tunnels of prairie dog towns. He imagined the rumbling hooves of bison herds with no end. He heard the sharp cry of red-tailed hawks and the sweet whistle of meadowlarks. He splashed through the shallow waters of prairie rivers alive with trout and crawdads.

Peering from the train, Richard was surprised to find cattle grazing and corn growing tight to the tracks spanning the Colorado lowlands. As the sun set on the sheep and alfalfa of Nebraska, surprise gave way to bewilderment. No one, not the author of his favorite book, nor his parents, teachers, or friends had ever told Richard the prairie was gone. When the sun rose to reveal endless rows of Iowa corn, bewilderment gave way to heartache.

The sorrow over leaving the house in Marshall was tempered by the truth that the house was still there. Although unlikely, it was still possible his parents could change their minds, reopen the Nelson General Store, still possible he could return to the ponds and fields.

The sadness that filled the train car was a new brand of grief. It had no edge, no shore, no way out. Not a single buffalo. No head-high prairie grasses. No chance to return to something no longer there.

Richard didn't mention the loss to his parents when the train pulled into Madison. Not a word of it to his friends eager for a game of football. He lived alone with the awareness of all that was no longer on the Great Plains.

For the remaining years of high school, Richard dutifully attended classes but rarely did the work. As classmates filed forward to hand in homework, he'd sit tight. His poorest grades came in English. After he'd earned three Ds and an F in a writing course, the teacher, Mrs. Pellete, wrote, "Richard really needs to work on his grammar. He didn't complete several important assignments this quarter. Also, he persists in reading his book (reptiles, of course!) during class even after many warnings."

As soon as he got his license, he borrowed his parents' car and explored on his own. A hundred miles out of the city, he followed dirt roads until they dwindled into the weed-choked driveway of an abandoned farm or a boarded-up home. He peered under old planks, snake hook at the ready. He waded creeks and ponds, long-handled net in hand. He spent sweltering summer nights sleeping on the back seat, windows rolled tight against swarming mosquitoes. Fall and winter, he shivered through frigid nights in a too-thin sleeping bag.

Whether goose-bumped or bug-bitten, Richard began asking questions that would, for the rest of his life, step him into an ever-increasing intimacy with the natural world. He discovered that five-lined skinks scratched beneath oak leaves in dry forest openings. He realized red-bellied snakes were drawn to warm sun on cool mornings. He learned to distinguish the rattling roll of a mink frog from the popping chirp of a boreal chorus frog. Each solved riddle sprouted a fresh question. Each captured creature propelled him to seek the next marvel, lurking somewhere in the rolling hills.

The career counselor, prior to graduation, didn't know about those outings. He scanned through four years of Cs, Ds, and Fs and recommended

Richard enlist. "You'll make good friends in the army," he said. "The military will create great structure in your life."

That counselor didn't know Richard kept a journal he never shared with his teachers. He didn't know the young man before him, alone at night in the north woods, had already written this:

The last grey light of day casts a dull glow upon the silent leaves and trunks. High above, unseen in the clear but darkening sky, the sharp buzzing call of a nighthawk foretells the appearance of the evening star. A thousand frog voices emerge from the darkening expanse of a swamp, filling the humid air with pulsating rhythms. The shrill song of the toads fades and is replaced by short, conversing peeps of tree frogs mingling with exchanging croaks of wood frogs, each group, in its turn, trying its song out on the light evening breeze. Strange sounds, too, drift through the faded maze of trees and leaves; a short rustle here, a dull thud there, each distinct and mysterious.

But above all the sounds of the forest this evening there is one most beautiful, the song of the hermit thrush. Somewhere in the darkness he stands alone and sings for all to hear, and indeed they do hear for all is quiet and hushed when the last note drifts away. God created no other song of such magnificence, and in quiet reverence the frogs, the nighthawks, the whole forest listens. His song starts loud, high, and descends in a series of bursts of chiming melody which seem to issue from a thousand voices together, singing, fading . . . slower . . . fading. Echo . . . night.

STROKE OF A PADDLE

Richard was placed on academic probation his freshman year at the University of Wisconsin. Even under threat of expulsion, he couldn't muster interest in the required English, math, and philosophy classes. He slogged through that first year and then enrolled in summer courses at the University of Arizona, eager to return to the world of reptiles. In the desert, Richard found classes aligned with his interests.

Dear Mom and Dad,

They are letting me take a senior-level course on the natural history of the higher invertebrates. I just love it so far. It is all about the wild birds and mammals of the southwest, how they live, where they live etc. There is one junior, one senior, and all the rest are grad students except me. I will sure have to work for a good grade. I think I will love working at it though.

Long hours in the library came easy; the pages of the natural history text, like a great novel, seemed to turn themselves. No one was more surprised than Richard when he set the curve on the midsession exam.

Dear Mom and Dad,

I got the highest grade in the class. I sure was glad to see that. I guess Mom doesn't have to worry and fret about my grades here.

My dorm room is like a zoo. Some of my pets are: a big and medium bull snake, a western shovel-nose snake, a Sonoran ground snake, a mountain king snake, two collared lizards, two scaly lizards, two side-blotched lizards, two horned lizards, a tree frog, a big tarantula, and several non-deadly scorpions.

Come fall, Richard released his pets, returned to Wisconsin, and declared a major in zoology, set on a career in herpetology. On the Madison campus, Richard met an animated Scotsman by the name of Ken Taylor. Ken was a graduate student studying under Dr. William Laughlin, an anthropologist focused on Inuit culture.

In his thick accent, Ken talked with great passion of his year apprenticed with the seal hunters of Illorsuit, Greenland. He lamented that the art of kayak construction was dying with the old masters. He spoke of paddling with the last men still harpooning seals from the sleek boats in the ice-choked fjords. Richard listened, enthralled with the knowledge residing, not just in Ken's mind, but in his hands and senses. Here was a man who blended science with adventure, knowledge with action, facts with skills. A man capturing a fading culture with his mind and body.

Ken invited Richard to check out the Inuit-designed kayak he was building in his garage. Richard soon set up sawhorses and started a boat of his own. In between campus lectures and study sessions, the two friends cut and lashed boards. The skeletons of their seagoing boats slowly took shape in the landlocked garage.

Richard's enthusiasm for kayaks spilled into schoolwork. He slowly worked his way out of academic probation, invigorated by professors like Dr. Hugh Iltis, a shaggy-haired Czechoslovakian who taught biogeography of plants. A passionate conservationist, Iltis inspired Richard to give voice to all he had not seen from the train window on his ride across the Great Plains.

For the first time, he applied himself in an English class, finally sensing a purpose in the craft of writing. He submitted an essay called "The Forgotten Prairie" to his English professor, but it was really intended for Dr. Iltis. The opening paragraphs read:

There was a day when the wind blew free and strong off the Gulf of Mexico onto the prairies. These vast grasslands offered no trees or mountains to stop the breeze and blew from what is now Texas northward for 2,000 miles to the Arctic forests of Alberta. The wind whipped the mighty land of grass into undulating oceans of green. It blew into the eyes of a million buffalo and grizzled the hair of the antelope. It whistled and whirled through 1,000 prairie dog towns and rocked meadowlarks on their slender perches. From the land of Blackfoot, Hidatsa and Pawnee to the domain of the Cheyenne, Kiowa and Comanche the wind blew and the sun shone above. This was the prairie.

The ocean breezes carried the white man to the new land, and it was all to change. A million square miles of grassland became a million square miles of checkerboard grain fields and grazing lands. The wind erased the last trace of the wandering buffalo and the native Indian who followed it relentlessly. Today, the fences of zoos confine the former denizens of the grassland and paved highways pass through the sterile lands of the Indian reservations. The plants of the prairie today can be seen only in forgotten corners of grain fields, along unused roadsides, or in a tiny tract of a university arboretum. The wind still blows free, but it finds a different land today. The prairies are gone.

The writing earned Richard his first A in an English class since elementary school. The essay also caught the eye of a classmate, who took it upon himself to mail a copy to then–secretary of the interior Stewart Udall. Richard knew nothing of this correspondence until a personal note, typed on embossed US letterhead, arrived in the mail.

January 30, 1962

Dear Mr. Nelson:

A friend of yours, Mr. Cameron Wilson, thoughtfully sent me a copy of your theme, "The Forgotten Prairie." Let me say that I appreciate not only the fine writing that went into it, but I am equally impressed with the fact that you chose this interesting subject. You have compressed a strong argument for the proposed Prairie National Park into your two-page theme, and I am grateful that you have put your constructive thoughts down on paper.

I sincerely hope that this item gets wide circulation: I am sure it will help in our efforts to establish a Prairie National Park.

Sincerely yours,
Stewart L. Udall
Secretary of the Interior

While Udall was in Washington, DC, trying to create a new national park, Richard and Ken were in the garage, stretching canvas skins over their kayaks. As they worked, Ken talked of the paddling prowess of the Greenland Inuit, how the men practiced rolling their boats in case they lost a paddle at sea. The best hunters, Ken explained, were adept at the paddle-free roll, cupping their bare hands to push themselves upright. Only an elite few could roll with clenched fists, holding a pebble in each hand to prove they had not opened their hands underwater.

Ken and Richard launched their boats in early spring, slipping the sleek crafts in between the rowing shells and sailboats tethered along the university's waterfront. They wiggled into the tight cockpits, picked up the hand-hewn paddles, and pushed off into the calm waters of Lake Mendota. Spring sun hot on his shoulders, Richard didn't know he would leave his beloved reptiles behind. He couldn't imagine he'd soon find himself amid vast herds of Arctic caribou, a northern echo of the vanished buffalo.

The two friends returned to the lake weekends and evenings. With Ken's guidance, Richard soon mastered the art of rolling. Together, they traded paddles for small boards gripped in their hands. Once proficient with the boards, they tried their open palms. Ken never quite managed the paddle-free roll, but he celebrated when his more athletic friend popped to the surface, arms overhead in triumph.

Holding a pebble in each hand, Richard then spent hours attempting the elusive close-fisted roll. He never surfaced with those stones, but he did emerge with a growing fascination with the men who could. He channeled his enthusiasm for the skills and lives of Inuit seal catchers into an exhaustive term paper for an anthropology course, which earned him an A from Professor Laughlin.

With most all his required classes out of the way, he negotiated an independent study with Dr. Laughlin. As an extension of his previous paper on the Greenland Inuit, Richard agreed to do a literature review on the wider topic of northern people's relationship to ice.

Lake Mendota was frozen solid when, in December of 1963, Richard completed the last of his classes, receiving his bachelor's in science just before the Christmas break. Richard had earned straight As and the praise of Dr. Laughlin for his seventy-page report, *Eskimo Knowledge of the Sea Ice Environment*.

New Year's Day, Richard received word that Dr. Laughlin wanted to see him. He was stunned to find himself in his professor's office listening to a job offer.

"You'd live and travel with the Eskimos, learn everything you can about ice and hunting," Laughlin said. "Your reports will be the backbone for a survival manual for air force pilots flying in polar regions. You'd be on your own for a full winter. There's nothing easy about this job. You'll be lonely. But you'll see things few people ever will."

Richard never figured out why he was chosen, why Ken Taylor or any number of more experienced anthropologists didn't get the job. Richard was untested, a potential risk to Laughlin's national reputation. But the professor saw something that made him roll the dice: perhaps a rare mixture of confidence and humility, creativity and discipline, gregariousness and humor.

And Richard's sense of adventure didn't allow him to say no.

KUSIQ

The steam bark *Beluga*, under the command of Hartson Bodfish, sailed from San Francisco in the spring of 1900 and, weeks later, dropped anchor off Point Hope, a village perched on the divide between the Bering and Chukchi Seas. Captain Bodfish went ashore to hire local men to help with the arduous work of killing and butchering whales. He also brought aboard Lucy Kongona, a Point Hope Iñupiaq, to sew cold weather gear for the crew. The *Beluga* chased whales all that summer, sailing northeast past Point Barrow and anchoring off Canada's Baillie Island just before freezeup.

The crew wrapped the ship's wooden hull with its canvas sails to conserve every bit of heat through the long winter. As soon as the spring thaw allowed, Captain Bodfish and his crew turned back to the hunt, eager to work after the idle months. The captain kept a daily log, chronicling the weather, the crew's morale, and animals seen and killed.

His entry on October 20, 1901, states he "paid and discharged the natives" at Teller, a settlement a hundred miles north of Nome. The following day, the *Beluga* steamed for San Francisco with its cargo of whale oil.

Five months after leaving the *Beluga*, Lucy Kongona gave birth to a baby boy named Kusiq Bodfish. Given the captain's marriage to a woman back in New Bedford, Massachusetts, it's no surprise his logbook makes no mention of his relationship with Lucy.

On subsequent trips to the Arctic, however, Captain Bodfish dropped in to check on Lucy and the boy bearing his name, delivering groceries and other big-city marvels. When Kusiq became school-age, the captain tried to take his child back to Boston for an East Coast education. But

Lucy, wanting her son to be a traditionally educated Eskimo, would not let him go.

When Richard arrived in Wainwright in 1964, Kusiq, also known as Waldo, was sixty-two years old, and one of the best hunters in the village. Like Richard's journey to the Arctic, his relationship with the great hunter was launched by a kayak. It was a windy September afternoon, spitting snow, when Richard assembled the pieces of the folding boat he'd hauled from Madison. Within hours, news of the white man's collapsible *qayaq* with a removable skin buzzed through town.

Kusiq was the first to stop by and check out the folding kayak. He was a big man, standing over six feet, his face creased by years of wind. His handshake was solid, his hand thick and calloused. Kusiq studied Richard's boat. He handled the wooden ribs, tugged on the canvas skin, fiddled with the metal latches.

The two chatted as a snow squall blew through, trailing a tattered patch of blue sky. Flickering grins bridged the distant worlds of the old hunter and young student. Before leaving, Kusiq bent down, gave the folding boat a tap, and mentioned he and his friend Ikaaq were working on a boat of their own.

> *In the afternoon I helped Waldo Bodfish [Kusiq] and Wesley Ekak [Ikaaq] as they prepared two ugruk [bearded seal] skins for an umiaqaluraq, or small, one man umiaq. The skin is laid outside, with the fat only partially scraped off. It is left outside for a week or so and then the hair can very easily be stripped off by hand. Following this the inside is scraped somewhat cleaner and is ready to be sewed. Waldo's wife Mattie sewed the two hides together after they were cut, using sinew (braided) thread.*
>
> *While all of this was going on Waldo, Wesley and I talked.*

Kusiq was a friendly man, generous with his knowledge. Perhaps his patience with an anthropologist's endless questions grew from his own mingled history with the white man's world. Ikaaq, too, who Richard described as "a very old man who enjoys holding school and teaching about hunting and the Eskimo language," proved a natural tutor. The two hunters, dear friends, told stories and laughed as they worked.

As the men stretched the sealskin over the wooden frame, the sky filled with the voices and wing whistles of birds heading south in advance of the coming cold. Richard asked about them all. He learned oldsquaw ducks were called *ahaaliq*, in mimicry of their ceaseless call. The four species of eiders, favored for their taste, were collectively referred to as *kaugak*.

Kusiq Bodfish working on an umiaqaluraq (Nelson archive photo)

Ikaaq talked of boyhood days living in sod homes heated with seal-oil lamps. He described hunting ducks with a *qilyamitaun*, a bola consisting of half a dozen walrus teeth, each attached to a string of braided sinew. Ikaaq said it was best to hide on top of a hummock in the flight path. With luck, the flocks would be close enough for the hunter to tangle a bird with a toss of the spinning weights.

It was Mattie who, after days of listening to Richard's questions and the old men's stories, said, "We're going to call you Niglik. You want to know so much about birds."

Niglik, Mattie explained, was the name of a small goose with a white necklace. When Richard opened his Peterson bird guide, Mattie pointed to a brant and said, "That's the one. That's who you are."

A mere eight days in the village, and he'd been honored with an Iñupiaq name and a fledgling relationship with the village's greatest hunters. Richard might have fully embraced being Niglik if he weren't preoccupied with becoming an anthropologist. He might have celebrated Kusiq's initial friendliness were it not overshadowed by Professor Laughlin's expectations.

Richard had a sea ice report due at the end of each month. With open water lapping the beach and the pack ice a mile offshore, his only access to ice was through the minds of the other men.

At Kusiq's invitation, Richard visited the town's coffee shop—a simple one-room affair, no chairs, just a table for the coffeepot. Evenings, men gathered and sat on the floor, legs stretched out straight. Richard leaned against a wall and listened to the talk of caribou and whitefish, foxes and ducks, ice and bears, wind and whales. Listening is an art, and Richard was a gifted yet impatient practitioner.

> *Whenever a visitor comes [to my house] it would be completely quiet if I did not talk and start conversations, and it sometimes becomes rather difficult to guide conversations for a long time. The same is true at the coffee shop, where I go each night. If I do not talk, they will speak in Eskimo, and if I do they will give replies, but never good ones.*
>
> *One of the difficulties of learning through conversation is that the men will very often answer a question by saying "wait till freeze up and you'll see." They don't wish to talk about it but will tell you when the time comes you will be told. This attitude of living in the present, not the future, is very typical, but does not help me at all.*

Richard shared his frustrations in letters to Ken Taylor, who was studying and living on Kodiak Island. Ken wrote back, warning his friend not to ask too many questions, to allow several weeks just to become familiar with folks. "Then," wrote Ken, "once an absolute maximum of the questions in your mind have been answered by observation rather than by ignorant, irritating questions, you might proceed to formal inquiry to clarify, quantify, and elaborate what you already know."

One evening, Burrell Negovanna, an Inuit man about Richard's age, stopped by. Richard served tea. Between sips, Burrell pulled a small notebook from his breast pocket, scribbled a few words, and then tucked it away. Tea nearly gone, Richard asked what he was up to.

"Oh, nothing," Burrell replied, "just writing things down in a notebook that no one ever gets to see."

Point made. For the rest of his career, Richard never took notes in front of people. Alone at night, he'd quickly outline the day's lessons then spend hours pulling details from memory. He'd also leave his journals open and available to anyone who dropped by his house.

He quickly embraced the wisdom of Ken's advice as well, as he wrote in his journal in October:

> *I am continually impressed that I was too inquisitive the first month I was here, and have now gone along with whatever conversation they keep up. Eventually, they always get around to hunting.*

Richard's nearest neighbor was Tagruk, one of Kusiq and Mattie's thirteen children. Tagruk inherited his father's friendly nature and took to trudging across the windswept snow almost daily to visit with the new white man. He filled many evenings with tales of wounded bears, wary seals, sudden storms, and shifting ice. As soon as Tagruk headed for home, Richard scratched the stories onto the pages of his thickening journal.

> *Wayne Bodfish visited this evening. He said that he once shot a bear, aiming properly for the top part of the neck. But he hit too low in the neck and the bear charged. He waited until it got close, he said, and then shot it in the body above the leg. He said you have to know where to shoot to stop a bear, by breaking its bones.*
>
> *He said that when retrieving seals you always should have an umiaqaluraq, and that he never goes out without one. When towing a seal he says to hold the line between your teeth. Don't attach it to your boat. This is because if a walrus comes to eat the shot seal you can quickly let it go.*

Unlike some of his brothers, Tagruk had never left the village to attend "outside" schools. Like his father, he was proud to be a traditionally educated Eskimo. When Richard created a stack of flash cards to learn Iñupiaq words, Tagruk eagerly helped with pronunciation and spelling.

Their long evenings together filled with a potent mix of curiosity that pushed both ways, one culture into the other. Before heading back to his own house, Tagruk often sang a little song he'd composed for his new friend. The song hinged on Richard's Eskimo name, Niglik. Tagruk's lyrics aimed to tease, but the short melody often ended with a chuckle of affection.

> *Nigligaichguk tunnguraalarut, tunngiich aasii kangiqsilyaangmigaich.*
>
> The brant goose speaks in English, but the white man doesn't understand.

CRONKITE AND CARIBOU

Wainwright had no phones, no televisions, no snow machines, no retirement plans, no checkbooks. A man's wealth was calculated by the speed of his dogs and the size of his meat cache. A man's education was gauged by the precision of his predictions about when the ice would crack and the caribou would come.

Such elegant ease was a world apart from the tension and buzz of US politics and culture. In the spring of 1964, the Beatles held the top five spots on the Billboard chart, the first Ford Mustang rolled off the assembly line, a dozen young men burned their draft cards in Central Park, and microphones were found embedded in the walls of the US embassy in Moscow.

That summer, President Lyndon Johnson, pressured to respond to brutal killings by the KKK, signed the Civil Rights Act. *Mary Poppins* hit the theaters. Three days of race riots left lingering fires throughout Philadelphia. Each evening, Walter Cronkite summarized the tumultuous times and political pressures, none of which made a bit of difference for the caribou nibbling lichens on the outskirts of Wainwright.

In early September, a cluster of folks gathered in the warm sun beaming into the White House Rose Garden. Conservationists Mardy Murie and Stewart Udall were among the tight bundle of officials standing behind President Johnson as he signed the Wilderness Act into law. Many of Richard's classmates at the University of Wisconsin celebrated. They cheered the persistent voices calling for restraint in the face of relentless sprawl.

The landmark legislation went unnoticed on the tundra. For men like Kusiq and Ikaaq, living where the sea could open and swallow all trace of a hunter, where wind-drifted snow often covered tracks as soon as they were made, a law creating places "untrammeled by man" made no sense. The wind gusting across the tundra on September 3 made no mention of the accomplishments of those gathered in Washington, DC.

> *West wind blew strongly all day, about 15–20 mph, kicking up a fairly heavy surf which broke in the shallow water 50 yards offshore from Wainwright. Skies were mostly overcast, the occasional snow flurries or snow showers passed. The sea was clear of ice, and the pack was over the horizon.*
>
> *About 12:30 the Wein plane came in on its regular Thursday run and landed as usual south of the village on the beach. The people generally wait in a large group for the mail to be passed out from the small post office building.*

That regular Thursday plane brought in boxes of candy, loaves of bread, and cans of soda. The same plane carried children out to be reeducated in the schools and cultures of faraway cities. Despite these changes, the psyches of the old men centered, as they had for thousands of years, on the differing habits of ringed and bearded seals, the movement of wounded whales, the crack of shifting ice. For the old hunters, lessons rising from the Paleolithic carried more relevance than any presidential proclamation. These men had never heard of Cronkite but knew more about caribou than anyone alive.

Despite his immersion in Wainwright life, Richard hungered for an afternoon game of football with friends ready to laugh at his jokes. Three thousand miles away, Flo packaged up fresh cookies and warm socks for her homesick son. Bob placed a reel-to-reel tape player in front of the television and recorded the weekly Packers games.

Richard chomped through the cookies and listened to the games before threading the tapes through his own reel-to-reel machine and recording stories to send back to his folks.

The tapes, saved by his parents, reveal a man caught between his passion for the new and his pining for the familiar. Whether he was whining about cooking or exclaiming about the northern lights, it must have been a great relief for Bob and Flo to hear their son's voice from so far away.

Richard tried to persuade his friend Roger Poppe, a grad student in Madison, to join him in Wainwright. In a letter recorded in early October, Richard told his parents of his wishes:

> *Dear Ma and Pa,*
>
> *I haven't heard from Rog yet whether he's interested in coming up. At times, I would offer him my whole salary to do it. Believe me, I almost pulled the plug last week, thought I absolutely had had enough. This business of living alone gets me down pretty low at times.*

Richard well may have pulled the plug were it not for his friendship with Ray and Barbara Bane. Originally from West Virginia, the Banes promised family and friends they'd teach in Alaska just one year. The young couple traveled north and never came back. After a stint teaching in Sitka, the Banes, hungry for wilder ground, accepted teaching jobs in Utqiagvik and then Wainwright.

Ray and Barbara had been there for two years when Ray looked out the window and saw what he described as a "lanky fellow walking—almost bouncing—across the village." The newcomer stopped by the school to introduce himself, and later, over dinner, explained his intent to document Iñupiaq ice survival skills. In his journal that night, Ray wrote: "He seems like a babe in the woods. He is quite young and enthusiastic about everything. Hope he makes it."

In the following weeks, Ray helped Richard assemble a team of dogs. "You have to be able to take care of yourself," Ray said. "You'll get no respect, no invitations to hunt, unless you can travel on your own."

With Ray's help, Richard bought four older dogs for seventy-five dollars and a young, vital one for an additional twenty-five dollars. After sketching and measuring a variety of sleds, Richard purchased wood and built a sled of his own. In a letter recorded on a late October evening, Richard described the huskies staked outside his door.

> *I made my own harnesses and spliced up a tow line. When you're going to go out mushing you put a hook in the ground behind your sled and then string up your dogs. Then you get on the back of the*

sled, pull up that anchor and hang on for dear life. I tell you it's a real ride. The tundra is flat but bumpy and those dogs really go. Greatest sport I have ever found.

Ray took his new friend on a caribou hunt a day's sled ride east of town. When a small herd wandered within range, Richard took aim at the lead animal.

The bull fell.

He'd later learn to select the cows, fatter and better tasting than the large males. But returning to the village with his first caribou strapped to his hand-built sled pulled by his very own dogs, Richard indulged in a rare moment of pride. Since arriving in the village he had been overwhelmed by all he didn't know about Arctic living. But that evening, blood staining his hands, whisper of runners on snow, Richard was elated by all there was yet to learn.

He hired a local seamstress to transform the caribou hide into fur pants, a fur parka, and a pair of fur mukluks, called *tuttuliks*, which he described in another letter home.

Caribou clothing is the warmest insulating material known. Down is almost as good but bulky. Caribou is very light, it's almost weightless. The only problem is that, under normal use, it only lasts about a year because the hair sheds so fast. Turns out you're eating hairs about half the time. But, Mom, at least you don't have to worry about me freezing.

Richard's warm clothes did little to guard against the chill of some households. In contrast to Kusiq's and Tagruk's welcoming natures, other families remained distant, wary. The children would stay back, sticking close to their parents and staring at the stranger. As Richard learned more of the language, he realized some parents were giving their children this warning: "*Iidigi. Tunnik.*" ("Danger. White man.") It was an understandable warning; white men had brought many unwelcome changes to the Arctic.

Understandable, but to Richard still painful.

RICHARD'S MIND OPENED TO THE new sounds and complex grammar of Iñupiaq at an astounding rate. In just a few months, the gibberish of coffeehouse conversations revealed broad shapes of context and then, bit by bit, the fine texture of details. "Richard gained a greater fluency in three months than I achieved in three years," Ray Bane said. "He is a gifted linguist—a genius really.

Despite Richard's knack for language, he found Iñupiaq humor elusive. When he once joined the levity with a chuckle of his own, a man briefly switched to English and snapped, "What are you laughing at? You don't understand." Another time, a hunter asked a question in rapid Iñupiaq, and when Richard was unable to answer, he said, "Ahh, you never learn."

It would have been easy to avoid the coffee shop and pass every evening in the warm oasis of the Banes' house: home-cooked meals, light and familiar conversation, relaxed and steady humor. Yet Richard limited himself to a single evening per week with Ray and Barbara. He couldn't learn from the hunters if he wasn't with them. And if he wasn't learning, he'd have nothing to report to Professor Laughlin.

The Arctic wind shook Richard's little house and set up a lonesome, unrelenting whistle as he wrote another letter home.

Dear Mom and Dad,

Today it was about zero all day, and the wind is a howler from the east, making it pretty darn brisk out. Stayed inside all day, even though this lonely house is virtually driving me out of my mind. The Eskimos are very hard on white men and give them credit for nothing. I have had a rough time, and being all alone I have nobody to talk to or to be with me.

The same month, Richard tethered his dog team to his newly built sled and traveled with the other men to an outcrop of coal, the town's main source of heat. Richard took his turn with the pickax, chipping at the black rock. He then helped sack the loose coal, stacking the bulging bags along the mine's entrance. At day's end, faces and parkas smudged with black, the men raced back to the village, Richard and his small team straggling in a half hour behind the faster sleds.

Other days, Richard drove his dogs to a freshwater lake three miles south of town to cut ice. Like everyone else, he had no faucet in his little house. He explained in his journal that each drop of fresh water came from cakes of ice placed in a garbage can warmed by the heat of his home.

Today I cut ice with Homer and Dempsey Bodfish at the village ice lake south of here. The people use very large saws, perhaps 6 feet long, and cut the blocks. An ice chisel, made of steel and pipe, is also used for part of the cutting. The ice was about 7" thick. Working in pairs, the men can get 120 to 160 slabs (2' x 4') in a day.

Today, David did all the cutting and Homer and I hauled the blocks up to the shore, where they are stacked on end in long rows. Each family or household has a stack, which is marked with their name. Very little is hauled back to town now because the trail is very rough and the ice gets broken.

They say 25–60 slabs will serve me for the year, but I will probably shoot for a larger figure. Of course the slabs must be stored in ice cellars in the summer.

In addition to sacking coal and cutting ice, the men hunted the caribou wandering past the village. Skinned haunches crowded the town's meat racks. Talk of caribou pushed all other subjects from coffeehouse conversation. Often, before leaving the coffee shop, a hunter might say: "I hear there are caribou near the Kuk River. Maybe tomorrow morning, I'll go."

Unsure if these oblique references were actual invitations, Richard loaded his sled and followed a group of hunters on a late October morning.

At 10:30 am I departed Wainwright alone. Temperature was near zero and fell to –5 degrees by nightfall. I took my five dogs and small 8½ foot sled. The men with whom I later camped had left earlier, most by 9:00 am.

On the tundra the trail was difficult because of the deep soft snow in the depressions and the uneven nature of the terrain. I quickly crossed over to the river, and found the trail to be quite good. The recent high winds had swept away much of the snow, leaving only an inch in most places, with some deeper but hard-packed drifts.

I followed the trail of the men up the Kuk River and up the Omalik River. As I reached a point where they all turned off onto the tundra, I caused a group of caribou to start running and soon heard shooting. I was able to find the men quite easily then. They shot six of this herd of caribou, which they stalked on foot and had taken four earlier in the day.

They quickly unloaded their sleds and went to pick up the caribou, returning as it was getting quite dark. We then staked out our dogs for the night and skinned the caribou. We set up two tents and I cooked a meal of boiled caribou tongue and heart.

That night, the men shared their meal within the cramped confines of a canvas tent, a gentle heat radiating from a hissing stove and a glowing lantern. A layer of caribou skins covered the snow in the floorless tent. Richard sat shoulder to shoulder with the other men. The men gripped lumps of meat with their teeth, cutting off bite-sized chunks with a deft upward flick of their knives. Richard followed suit, cautious not to slice the tip of his nose.

The next morning was clear and cold, –10 degrees, with patches of ground fog. The men broke camp and traveled inland to a pile of butchered caribou stashed on a previous trip. While the men loaded frozen carcasses onto their sleds, one of Richard's dogs chewed through its harness. Richard pulled out his sewing kit and set to work. The other hunters lifted their snow hooks and left.

Every few minutes, Richard set down his needle, drew his arm inside his parka and defrosted his fingers in the warmth of his armpit. When the webbing was finally stitched, he pushed stiff hands into fur mittens and scanned the tundra. The other teams were nowhere to be seen.

It was midday. Three hours of daylight remained. The village lay twenty miles to the west.

Richard's team struggled with the heavy load. To ease their burden, he stepped off the runners and trotted beside the sled.

Near twilight, a breeze picked up and blew away the tracks of the other men. After dark, the winds lifted into a full gale. A curtain of blowing snow obscured everything but the stars directly overhead and the dogs. Following coffeehouse advice, Richard let the lead dog choose the way. Mile after mile, he tromped alongside his team, the wind gusting steady over his right shoulder the only indication they were traveling true.

A few dogs barked when they slid into town, but there was no light, no movement, nobody awake to make sure the new guy was still alive. He'd run a marathon in heavy clothes, exhausted to the edge of delirium. Richard staked his team and stumbled into bed, too tired to undress or unload his sled. If anyone was surprised or impressed to see his dogs in the morning, they didn't say.

In the following weeks, the men hunted caribou every day. The migration peaked in late November, with thousands of animals moving past the village. Richard estimated over four hundred caribou killed in a two-month period.

Richard cached his own pile of caribou. He fried steaks for himself, boiled batches of meat for his dogs, and, when he had time, shared stories with his folks.

> *Last night they had a dedication of the new Assembly of God church. Got home at 6:30 and the dedication was at 7:00. I had to take care of one caribou. It had blood all over it. I couldn't leave it on the ground because the puppies would come and eat on it. So, I tipped one of my sleds up-side-down and stuck a barrel underneath it. I grabbed that caribou and managed to hoist it on top of the sled. I came in the house and grabbed my song book and ran to church. I sat down and was singing away with my hymn book resting on my leg.*
>
> *When the song ended, I closed my book and saw it was covered with blood. I realized both of my thighs were covered with blood from the caribou. I inspected myself a little more and saw that I had blood all over the front of my coat and my boots were covered with blood and there was blood all over the floor where I was sitting. At home this would have been a tragedy beyond compare and I would have had to get up and run out of church. But this is Wainwright and probably helped my status in the village. Old man Ikaaq looked over at me and smiled as if to say, "Ah good, you got caribou."*

In late November, weeks of high winds and low clouds gave way to crisp, still air. Richard hitched his team and mushed up the Kuk River. Herds of caribou were scattered all around him, but he carried no rifle that day. He set the snow hook on the riverbank and watched the last sunrise of winter. "It never did anything but creep above the horizon," he later said of it. "Like a long sunset. It was beautiful flaming red, like a huge, great big pulsating fire."

He pushed back his parka hood to take it all in. Backlit breath from his dogs rose like golden smoke. Sky-wide silence. Frozen waves of tundra rolling in all directions. Slender, elegant caribou silhouetted against the glowing skyline, a reflection of the stocky bison he never got to see on the Great Plains.

The cold bit his ears and burned his cheeks. He pulled up his hood, lifted the snow hook, and turned his team toward home.

BORDERS

Saalguaq refers to a freshly frozen apron of ice. It's thin enough to shine black with the sea underneath. Seals can push through to get a breath. It bends and then breaks beneath the weight of a man.

It is here, at the border between predator and prey, that the genius of Richard's Iñupiaq neighbors comes fully into focus.

Every one of Richard's journal entries through the shortening days and deepening cold of fall makes note of the weather and the ice, as in this entry on the last day of September:

> *The skies were mostly clear with low temperatures around 12 degrees. Some slush ice formed along shore last night and remained all day today, increasing in distance from shore around evening.*

Two weeks later, Richard hooked up his dogs and traveled north.

> *Temperature 0 degrees in the morning, up to +4 in the day, and down to –8 at night. Along the shore north of Wainwright was little scattered pack ice chunks but a fair amount of young ice. From ten miles north up to Atanik the chunks of pack ice along shore got very thick and the young ice disappeared.*

After two months in the Arctic, Richard had yet to set foot on the frozen sea. He yearned to shadow the town's hunters as they navigated the treacherous terrain. He quizzed the men about what he would need for ice travel and compiled a list of essential items in his journal:

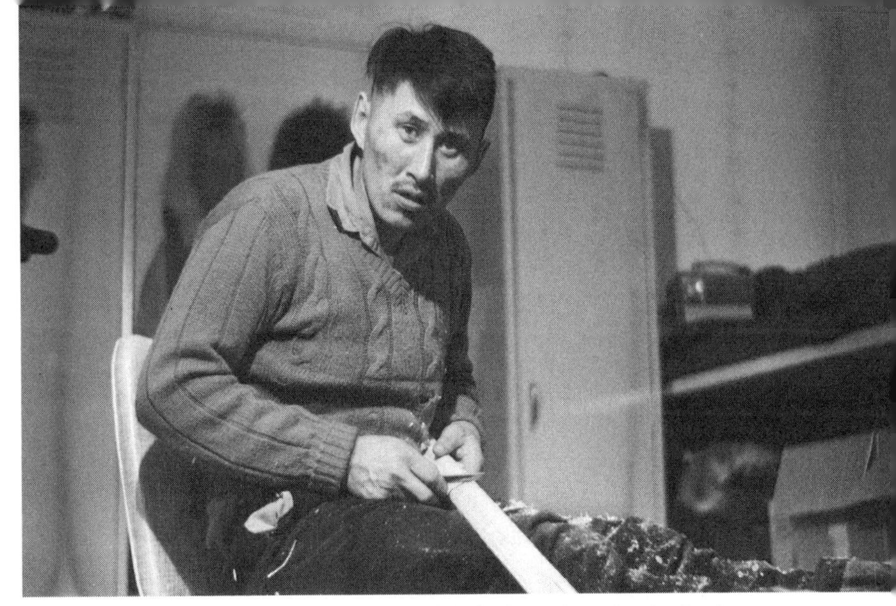

Tagruk Bodfish shaping a kayak paddle in Richard's house (Photo by Richard Nelson)

1. Unaaq (ice tester)
2. Manaq (floating seal hook)
3. Crescent axe
4. Saw (cutting snow blocks)
5. Large knife
6. Wolf mittens
7. Tuttuliks (caribou fur mukluks with ugruk covering)
8. Mukluks (ugruk soles)
9. Caribou socks
10. Fur pants (caribou, bear, or dog skin)
11. Caribou parka (inner and outer)
12. Fur wristlets (caribou or fox)
13. Harness for dragging killed seals
14. Fur muff (caribou)
15. File or stone (knife sharpener)
16. Tent (light, two man)
17. Caribou skin rugs
18. Shovel

Beneath the hissing glow of the Coleman lantern in Richard's one-room house, Tagruk helped create the needed tools. They began with an *unaaq*, a slender, seven- to eight-foot wooden staff with a metal point on one end and a curved hook on the other. Hunters probe young ice with an unaaq to evaluate its thickness. The hooked end is used to retrieve seals or pull a hunting companion onto thick ice should they fall through.

As they worked, Tagruk taught Richard the names for different types of ice. *Uqsrugiiraq*, the earliest stage of freezing, causes wind ripples to disappear from patches of the water surface. *Mugatłiq* is slush ice, the midstage between uqsrugiiraq and the dangerously thin saalguaq. Saalguaq thickens to become *sikuliaq mapturuaq*, which is gray in color. The thicker ice is strong enough to support a man, and can no longer be penetrated by a single sharp thrust of the unaaq.

Subsequent evenings, the two men crafted a *manaq*, fitting three metal hooks onto a pear-shaped chunk of wood with a long line attached. Tagruk described tossing the manaq just beyond a dead seal. With a quick jerk of his arms, he demonstrated the sharp tug needed to set a hook into the seal's skin. Hand over hand, he pantomimed the gentle touch required to then ease the body to the ice edge. Once the seal is snugged against the ice, the hooked end of the unaaq is worked into the bullet hole in the seal's head before dragging it across the thin ice.

As the days shrank and the sea ice thickened, Richard noticed that stories of ice travel also became the focus of coffeehouse conversation:

> *Tonight the men talked about walking on thin ice, mentioning there is a special way to do it and that some people are better than others. Freddy indicted that he was unable to stay up on ice less than about 2" thick, and Raymond can stay up on thinner ice but cannot do as well as his father. They say there was once a man who could walk on ice so thin that his toes would go through on every step but he would move along without going through. To stop would, of course, result in immediately breaking through.*

Old Man Ikaaq told Richard that on thin ice you've got to move like a bear, spreading your legs wide to distribute your weight. But even Ikaaq had once fallen through when he blundered into a patch of *mapsaaq*, open water or thin ice covered by drifted snow.

He tried to get back on thick ice but his gun (it was slung on his back crosswise) caught under the ice edge. He finally took out the gun and threw it on the ice then clambered up onto and rolled in the snow to beat out some of the moisture before going home.

Although cautious of falling in the Arctic sea, the men were even more wary of getting stranded on the seaward side of an opening fracture in the ice, called a lead. One evening, coffeehouse chatter centered on the story of a man's disappearance a long time ago. Assuming him dead, everyone had been surprised when, three years later, he walked back into the village and told his tale.

Unable to cross a widening lead, he'd built a snow house on drifting ice and survived for weeks on hunted seals. The ice pack carried him all the way to Siberia, where he settled in and learned the language of those far-distant neighbors. While there, he had an elaborate Siberian design tattooed between his shoulder blades. One late winter, on solid ice, he walked home, carrying the proof of his extraordinary journey on his back.

Richard's journal filled with tales of less fortunate men: young hunters found starved and frozen, men with legs crushed by colliding ice floes, people who drifted away and never came back.

Despite these stories, Richard was still eager to get on the ice. The other men, in contrast, patiently waited for a strong onshore wind to jam heavy pack ice against the coast, crushing it into massive ridges anchored to the ocean floor. Finally, it happened.

By late November, the entirety of the visible ocean was covered with jumbled sea ice. However, the wind died last night and there was not as much piling as some had expected. No open water was visible to me although Kusiq said he'd seen steam rising above an open lead.

Today was the first day any sea ice hunting was done. Greg Tagruk was successful in getting one ugruk.

People have warned me the ice will break away with an offshore wind as soon as the current stops flowing from the south. This is because the ice will lower and crack with the falling tides and the north or east wind would push it away.

Every day, Richard watched hunters head out onto the ice. Each evening, with a mix of envy and awe, he examined the seals they hauled back to the village. Hunkered in the coffee shop, he questioned the men about what they'd seen, the conditions of the ice, the movement of the seals, the equipment used.

One morning, curiosity overcoming caution, Richard awoke and walked onto the ice alone.

> *At about 10:30 am, as it was getting quite light outside, I went out to the lead on foot. I found the tracks, from yesterday and today, of hunters going out to the lead so I followed them. The ice was quite rough, mostly large chunks of ice piled together and large slabs of ice rafted over the other.*
>
> *At the edge of the lead there was young ice, extending 20 to 50 feet out from the heavy ice. I found the place where someone had shot a seal yesterday, and could tell from the blood where it had been pulled up over the ice edge. There was about 15 to 20 feet of new ice formation beyond the point which thus marked yesterday's ice edge.*

It was a cold December day, –22 degrees, with a light easterly wind. But decked from head to foot in caribou-skin clothing, Richard stayed toasty warm. With his pale face and blond hair tucked deep into the shadow of his fur ruff, he would, from a distance, have looked like an Iñupiaq. But, unlike his neighbors, he carried no rifle, no manaq, no tow strap to drag home a dead seal.

Richard would, soon enough, become obsessed with seal hunting, but on his first day exploring the frozen sea, the serene beauty of the ice pack was reward enough. He stood, for long minutes, mesmerized by the *puyugruaq*, or dense fog, rising from the black gap of open water. He took tentative steps onto the sikuliaq mapturuaq, testing the young ice with frequent jabs of his unaaq. He stared across the steaming water, scanning for the round head of a passing seal.

After an hour of wandering, he saw the dark silhouette of a hunter moving across the expanse of flat ice. The two men walked toward each other. As they drew near, the hunter pushed back the hood of his parka, and

Richard recognized the face of Simon Tazarook, a young man who frequented the coffee shop. Richard pushed back his own hood, and Simon nodded and said, "Oh, it's *you*."

News of Richard's solo venture quickly spread through town. Richard had been home just long enough to slip out of his parka and eat lunch when his neighbor stopped by. Tagruk made no mention of Richard's day on the ice. He did, however, share that one should always check the current by dropping a light-colored object through a crack or opening and watching as it sinks. If the object goes any direction except away from shore, it is safe to hunt. If it moves away from shore, however, you must get off the ice fast. Tagruk also mentioned that if ice floes drifting past you in a lead suddenly stop moving, it means you have begun moving yourself and you had better run for shore.

Later, at the coffee shop, the evening filled with more advice. Taqalaq, a senior hunter, said if you find yourself on the wrong side of an actively opening lead, run in the direction of the current until the crack narrows enough to cross it. Tagruk's brother David said to keep from panicking it's best not to look down when you find yourself on thin ice.

Richard was disappointed to awake the next morning to a southerly wind that closed the offshore lead. He knew the ice might remain inaccessible for weeks.

> *The conditions of the ice are not good for ice hunting, and the outlook is no better for hunting in January, so I feel that I might best serve the purpose of the study by leaving Wainwright after the month is over unless a change occurs.*

Such impatience had no place in a culture carved by the slow passing of seasons. The men filled their days scouting for bears, trapping foxes, and cutting a trail through the rough ice so they could mush their dogs to the lead they knew would reopen soon enough.

It was a bitter cold day in January, –34 degrees, when Richard joined the trail crew.

> *From noon until 3:00 pm, in spite of the very severe weather, David Bodfish, Jim Aveoganna, Ward Anasugak, and I worked chopping a*

> *"smooth" trail out toward the flat ice of the old lead. The work went fairly fast, using axes and one pick axe, although the trail is not what one would call smooth. However, they said that it is a good trail, and would be smoothed by the sleds traveling over it, as well as the blowing snow.*

A few days later, Tagruk and Richard stood together on the drifted snow between their homes. The dogs, staked all around them, were curled tight, noses tucked beneath tails in the bitter cold. Tagruk gestured across the ice. "Look there," he said. "Puyugruaq. The ice is open."

He explained that an expanse of dark water reflects gray on low clouds. Richard squinted to the horizon and noticed a faint dark smudge separating white ice from clear sky.

"Seals swimming out there," Tagruk said with a smile.

The two neighbors walked to the coffee shop, mukluks squeaking on the wind-packed snow. With the lead looming offshore, all talk that evening centered on seals. Anaqqaq, one of Ikaaq's sons, asked Richard to join him the next day. They made plans to use Richard's new sled pulled by Anaqqaq's ten dogs. That night, Richard readied his gear, coiling the line attached to his manaq, leaning his rifle alongside his unaaq.

In the morning, Richard slipped from his cozy sleeping bag and pried his mukluks from the frozen floor. In a thick sweater and wool hat, he stirred a cup of cocoa and ate a quick breakfast. After shrugging into the bulky warmth of his caribou-skin parka, he kicked at his front door to break the rim of ice which formed each night.

Outside, the cold cut the back of his throat and stung his cheeks. Bristling stars in a moonless sky cast just enough light for Richard to find his way to Anaqqaq's house.

Anaqqaq was outside, tossing lumps of meat to his dogs. They worked together harnessing the team, clipping each animal in turn to the towline. The dogs yipped and strained against their harnesses, eager to run.

With the team in place, Richard sat on the sled, and Anaqqaq took his place on the runners.

Anaqqaq lifted the snow anchor. The sled lurched into motion. With a few sharp commands he guided his team off the tundra and onto the ice.

The dogs fell silent. The whisper of the sled lifted into the cold stillness between stars.

They followed the freshly cut trail through the jumbled old ice, arriving at the flat expanse of young ice as the sky brightened from black to gray. After anchoring the sled, Richard and Anaqqaq gathered their gear and walked toward the fog-shrouded water. They stopped near the ice edge, stomping their boots to drive blood into their chilled feet.

The distant sun licked the southern sky, casting enough light for a few hours of hunting. The two men stood together and stared into the twilight glow reflected on the rippled water. That moment would, years later, form the heart of Richard's first book, *Hunters of the Northern Ice*:

> *The Eskimo . . . dropped to one knee, and started to scratch the ice with a knife he carried in a belt that encircled his parka. I watched him, baffled. "Seal down there!" he hissed softly, pointing into the cloud of steam before us. Squinting into the grayness, I could barely make out a round black silhouette, moving silently in the distance. A moment later, it vanished abruptly, like a broken bubble. Instead of giving up, the Eskimo jumped to his feet and ran closer to the ice edge, then quickly squatted on the frosted ice. He began the mysterious rhythmic scratching again. "Watch, close by this time," he said without stopping.*
>
> *A few minutes later the seal's black head bobbed up in the water not 30 yards away. The Eskimo still scratched the ice, until the animal stared toward us and rose higher in the water, captured by some irresistible curiosity about the noise and about the two figures on the ice.*

As Richard centered the seal's round head in the circle of his rifle scope, he did not think about the intense cold or his distant family or the lonely nights. He did not think how peculiar it was for a lizard-loving midwestern boy to be seated on the sea ice, frigid rifle stock pressed tight to his cheek. Holding his breath to steady his gun, he entered a timeless space suspended between hunger and celebration, a moment echoing back through the lives of hunters crouched on the African savanna or sneaking along the banks of European rivers or pursuing prey across the Asian taiga. A space loaded with more lessons than the largest lecture halls. A space both expansive and focused, fundamental and fleeting. A space at the heart of every Iñupiaq hunter.

> *The seal's head broke the surface again, this time only 20 yards away. It stared into the white silence. . . . The shot split the air like shattered crystal. Before us in the smoking haze, the low rounded back of the seal settled on the pulsing water, amid a slick of glassy oil spreading from its lacerated blubber.*

Anaqqaq eased closer to the open water, the thin ice bending beneath his feet. He laid the length of his unaaq before him and stood on it to distribute his weight. It took him several tosses of the manaq before the line landed over the seal's body. He hooked the seal with a sharp tug and then eased the lifeless animal to the ice edge. He pulled the limp seal from the water and stepped back to thicker ice.

The mist shimmered pink with the last light of the short Arctic day. Anaqqaq glanced at the seal and then flashed a smile at his companion. Niglik smiled back.

ICE AND LAUGHTER

Richard never witnessed a conversation that did not, in due time, circle back to hunting. Every exchange, whether in the coffee shop or on the trail, drifted to the whereabouts of seals or caribou, the conditions of ice, the habits of dogs, or the proximity of bears. Preparing for the dangers of Arctic travel, the men shared nuanced knowledge about invisible currents and coming storms, whiteout conditions and emergency camps.

Richard came to see the men's single-minded focus as the engine of a brilliant, churning intellectual machine. A machine set in motion in the sod houses and ice shelters of their ancestors. A machine combining individual observations to produce insights unavailable to any single hunter.

In Madison, Richard's obsession with snakes and frogs was a private affair, tolerated but not shared by the adults in his life. In Wainwright, his fascination with the natural world found a home, moving from the shadows of a hobby to the full light of essential knowledge. Every aspect of his training as a hunter was rich with intriguing details that could be immediately applied, unlike his boring math classes and English lessons.

There was no end to an Iñupiaq education, no diploma declaring it was done. Nor was there a formal process to learn the intricate lessons of Arctic survival. Young men—Richard included—were expected to absorb through keen observation and careful listening. Lessons well learned were met with silent approval. Mistakes unleashed a barrage of ridicule.

Out hunting seals with Tagruk one day, Richard blundered onto an apron of young ice. Tagruk shouted a warning just as Richard felt the ice bend beneath his weight. The coffeehouse lessons paid off, and Richard resisted the urge to stop and turn around, a move that would have plunged him into the sea. Instead, he kept moving, arcing back to thicker ice to rejoin Tagruk, who was doubled over with laughter.

Throughout the rest of the afternoon, Tagruk burst into giggles each time he recalled his neighbor's brush with death. And by the time Richard made it to the coffee shop that evening, the whole village was chuckling with the news.

Dear Ma and Pa,

I have to admit, these Eskimos are really clever people, but I wouldn't ever tell them because they spend too much time telling me how stupid I am. Most of them are really great people, but they sure test my patience sometimes.

The threat of frigid waters was tame compared to the pain of public humiliation. Richard never again ventured onto thin ice.

The teasing eased up a bit after Richard killed his first seal. He butchered the animal on the floor of his house. It was a small *natchiq*, or ringed seal, and it was black with no spots, an unusual coloration called a *magumnusiq*. With guidance from Anaqqaq, Richard slit the hide from throat to tail, then separated skin from blubber with long strokes of his ulu. An easy levity, born from the day's success, flowed through the room.

Richard had just finished skinning when Tagruk stopped by. Tagruk and Anaqqaq encouraged him to give away the meat. Richard related the experience in a recording to his folks:

The accepted procedure is for a hunter to save only a little bit of his first seal and to give the rest away, especially to the older people. The giving of my first seal was quite unexpected. I don't know that a white man has ever bothered to do this. They sure seem to like it an awful lot. Boy, they sure have talked a lot about my first seal and giving it away and all that.

Once, on an overnight hunting trip, Taqalaq told Richard that near Christmas, people would gather for days of traditional games—the knuckle hop, the finger pull, and the high kick. "When we get back home, you say *'Anaktugniaqtugut,'* he said. "It means 'We're all are going to play together.'"

Richard repeated the phrase, and Taqalaq laughed, saying, "Yeah, that's perfect," and while mushing back to the village, he insisted that Richard keep practicing. So Richard shared the saying with everyone he met, all to great laughter.

Later, at the coffee shop, it was revealed that *anaktugniaqugut* did, indeed, mean let's play together. Richard, however, had mispronounced the first *k* as a *q*, which changed the opening syllable from *anak* ("games") to *anaq* ("shit"). Taqalaq had been encouraging him to proudly exclaim, "We're are all going to shit together."

For Richard, embarrassment was overshadowed by relief for his growing acceptance into the world of Arctic hunters. The learning of language went both ways, as many of the men wanted to improve their English. This two-way exchange gave birth to the phrase "Richard needs to work on his *utchucation*." Pronounced oo-choo-KAY-shun and derived from the Iñupiaq word *utchuk* ("vagina") and the English word *education*, this phrase became a steady piece of wall-tent humor.

The linguistic jousting continued when Richard tweaked a single syllable within a common church phrase. *Quyanagniagaadin*, used by the passionate faithful, figuratively translates to "Thank you from the bottom of my heart." Richard bastardized the phrase to *Quyagnaqniagaadin*, meaning "Screw you from the bottom of my heart." The men collapsed with laughter.

Mirth, as much as anything, bridged the cultural chasm between this midwestern man and his Arctic companions. While anthropology remained an abstract, suspicious activity to the villagers, they understood Richard's curiosity. They appreciated his linguistic skills and admired his athleticism. The blood on his hands from hunting allowed them to overlook his odd habit of scribbling in his journal.

SINCE ARRIVING IN THE VILLAGE, Richard had received logistical support from the Arctic Research Lab in Utqiagvik. The established protocol was for Richard to mail shopping lists to folks at the lab, who packaged food and other supplies and shipped them on the weekly plane into Wainwright. The system worked fine throughout the fall, but by January, Richard quit sending the lists. Imagining the young student languishing in the village, the lab people simply guessed about the twenty-two-year-old's needs and kept shipping boxes.

But Richard's tastes had begun to change. He fried up a few of the beefsteaks they sent him, decided that caribou was tastier, and tossed the remaining steaks to his dogs. He devoured the fig bars and made steady use of the bread and peanut butter, but much of the other food he jammed alongside the untouched cans of beans and boxes of noodles in his cabinets.

His dogs' diet, like his own, had a distinctly local flavor.

> *Around here you feed your dogs what they call "dog soup." You take a big pot and put in water, Friskies and cornmeal. Then you dress it up, add some sort of meat to it.*
>
> *Tonight was the most exotic soup I've made yet. It contains seal guts, a seal head, seal blood, some caribou meat and the oil that comes out of rotten walrus blubber when it thaws.*
>
> *The basic food, which is used more than dog soup, is walrus hide and blubber with some meat attached which is usually pretty rotten. Sometime you might throw in a sea gull or ptarmigan, any kind of meat really.*

As Richard adopted the local diet, the community around him relaxed. After a midwinter church service, Billy Patkotak leaned over and asked Richard if he had enough food for his dogs. Richard replied that he was, for now, doing okay. "Well, as soon you start running low, come over to my house," Billy said. "I got lots of extra walrus and want to give you some." He recounted Billy's generosity in a letter to his parents.

> *This is another example of how these folks can be the finest people you've ever known. Generally, I'm enjoying myself more and more and I'm realizing, more and more, that I am going to miss this place an awful lot when I'm gone.*

TEXTURED TUNDRA

Ikaaq had gladly traded his bola for a shotgun. Kusiq had given up his seal-oil lamp for a gas lantern. All the hunters would, in coming years, replace their dogs with snow machines. Richard, in contrast, was hungry to learn of the abandoned tools and fading techniques. His journal filled with schematics of seal-fat stoves and baleen fox traps. He quizzed Ikaaq and Kusiq about spearing swimming caribou from kayaks and netting seals through ice cracks.

For months, he'd admired a skinless kayak frame with broken ribs languishing on someone's meat rack. The owner was unwilling to sell the neglected boat, so Richard and Tagruk decided to build a new one. The boat slowly took shape, dominating Richard's tiny house.

When the last baleen rib was lashed to the wooden frame, Richard recruited Old Man Ikaaq to stretch sealskins over the skeleton.

> *The outstanding thing is the smell of my house. Which is absolutely beyond belief. I have these two ugruk skins in the house right now.*
>
> *What they do after they catch an ugruk is they peg the skin outside half the summer until it gets really rotten. And then they fold it all up and freeze it. What I had to do was get a couple of these frozen ugruk skins and bring them into my house to thaw them out.*
>
> *You hang them from the ceiling and you strip the hair which comes off with your fingers because they are so rotten. When the hair comes off is when they really start to stink. They are an oily mass. The grease just oozes out of them. The way they stink is just*

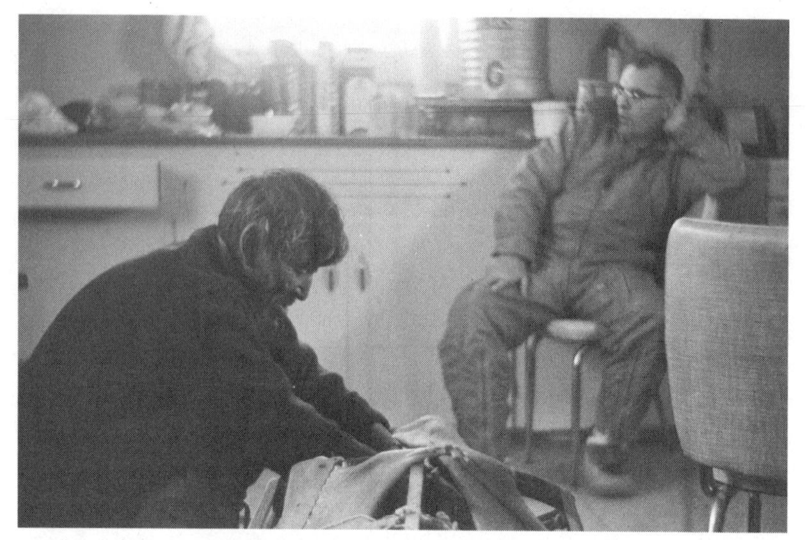

Wesley Ekak sews sealskin onto a newly constructed kayak in Richard's house. (Photo by Richard Nelson)

completely beyond description. The smell has permeated my clothes and my skin. You can't wash the smell off your hands, it just has to wear off. I'll have to burn my clothes at some point.

Tomorrow Ikaaq is going to sew it on the kayak. The kayak then needs to hang in the house for a week while the skins dry. So I guess I shouldn't get too bothered with the smell because it's going to be with me for a while.

While other men in the village had to balance their family obligations and home chores, Richard, living alone, was able not only to fill his house with an ungodly stench but also to spend more days searching for seals than any other hunter. As often as possible, he slipped into his caribou clothing, kicked the ice from around his door, and trudged onto the frozen sea.

Richard had read descriptions of killing seals with a harpoon as the animal surfaced at a breathing hole, or *allu*. Ikaaq and Kusiq could describe the hunting method, but even these old-timers now relied on the longer reach of rifles. Unable to find a mentor in Wainwright, Richard traveled to Utqiagvik to hunt with Pete Sovalik, an Iñupiaq man known to be well versed in the old practice.

Pete had killed a seal at an allu just days before Richard arrived and was happy to have the young *tanniq* ("white man") tag along on his next hunt. The two left Utqiagvik on foot, walking offshore a couple of miles before encountering the dome-shaped profile of a breathing hole. After situating two blocks of ice—one for a seat, the other for a footrest—upwind of the allu, Pete instructed Richard to walk around him in a wide circle to scare seals from any other holes in the area.

Pete didn't kill a seal that day, but Richard filled his journal with pages of detailed description and drawings.

Back in the village, he set up ice blocks downwind from promising-looking breathing holes. The seals never visited the allus where he sat vigil, yet hour after shivering hour, the animals swam through his imagination, careening through upended canyons of ice, chasing shadowy schools of fish.

While most memories become rounded by the relentless rub of time, the visions born in those hours of silent, expectant solitude condensed with frost-like clarity for Richard. A decade after leaving Wainwright, details tattooed in his mind allowed him to create this imagined account of an old Iñupiaq hunter waiting for a seal:

> *Young men said that breathing-hole hunting was too cold, that it involved too much waiting. The old men said only that people must eat. They had learned the art of enduring patience, as if they could merge their thoughts with the timeless physical world that surrounded them. . . .*
>
> *Sakiak was enveloped in still silence, interrupted only by the occasional buffeting of wind against his parka hood. His breath condensed on the ruff around his face and on his scraggly moustache, coating each hair with thick white frost. He could feel the immensity of the ice pack surrounding him, its quiet, latent power. . . .*
>
> *Sakiak drew his arms from the sleeves of his parka and held them against his body for warmth. He was shivering. Frost had collected on his eyelashes and brows. Occasionally he poked a bare hand up through the neck of his parka and held it against his cheek to warm the stiff, numb flesh. His toes felt large and icy cold. . . .*

A growing ache spread up Sakiak's legs and back, but he dared not move to relieve the discomfort. The seal might be near enough to hear any noise transmitted through the ice to the water below. . . .

He was shivering hard now, and he wondered if his shaking might jiggle the ice stool, making a noise that would scare away the seals. He smiled, thinking what a great joke that would be after such a long, cold wait!

But beneath him at that moment a seal torpedoed through the black-gray water, darting and arcing in pursuit of the fleeting silver of fishes. It dodged between the blue and emerald-green walls of ice protruding downward beneath the hummocks. Huge inverted ice mountains blocked its path, but it sensed them and turned away before striking invisible barriers deep in the blackness. . . .

For more than a minute the seal remained motionless, ignoring the fish that swam too near. It was in need of air and was listening. Then it suddenly whirled and shot upward toward a circle of white that glimmered faintly in the high distance. . . .

In the silence of the pack, after the long wait, the seal's approach was startling and exciting. Sakiak first heard, almost sensed without hearing, a pulsation of the water inside the allu. He then saw water flow through the opening and over the ice outside, where it instantly froze to a fresh glaze. This water was forced up ahead of the seal as it rose from below.

Sakiak heard scratching as the seal cleared away the newly formed ice at the tunnel's upper opening. He quickly slipped his arms into the sleeves of his parka, then remained perfectly still. The cold had vanished. Shivering ceased as warmth spread through from mind to muscle.

He fixed his eyes on the allu, consumed with intense concentration. His lips moved slightly, almost imperceptibly. "Come seal," he whispered, asking the animal to give itself to him. "Come. . . ." It was only a thought this time.

In a moment the seal obeyed Sakiak's will. It took a first short, hissing breath, smelling the air for signs of danger. He did not move. He expected the brief silence that followed, knowing the next breath would be a deep one.

Whoosh!

It was a long, drawn-out hiss that sent a misty spray from the opening. This noise was loud enough to drown out the sound of Sakiak's movement as he reached down and picked up his rifle from his legs. He was careful to spread his arms so his clothing would not scrape noisily, and he was still before the deep breath was finished.

Whoosh!

Again the animal breathed. Sakiak lifted his rifle and held it vertical, with the thumb of his upper hand against the trigger. Again he waited, as the second breath stopped.

Whoosh!

On the third breath he moved his rifle straight above the allu, its muzzle inches from the opening. His face was expressionless. His resolve was complete.

As winter opened further into spring and the light returned, Richard's time in the village dwindled. He spent longer days on the ice, staying up late into the night scratching out his field notes. Almost every letter home begins and ends with a declaration of how intensely busy and deeply tired he was by the self-imposed workload. The recorded letters, mostly made in the wee hours, reveal a voice tinged with fatigue, punctuated with yawns.

At the end of March, Richard began prepping for his return to Wisconsin.

It's getting warm, up to 30 degrees or thereabouts almost every day. And the days are really long now, with sunset around 9:00 pm or 9:15 pm. Everybody has suntan faces, but of course the rest is very white.

I am sending 3 or 4 boxes. There are fur things in them so please put them in a cool place. Some of it is sealskin, which really stinks, but it is in little plastic bags. At any rate, if it smells a little don't worry about it.

I am very ready to come home, although the thought of no more dog mushing and seal hunting is not a happy one.

Richard was scheduled to leave the village the same way he arrived; aboard the Cessna operated by the Arctic Research Lab in Utqiagvik.

Richard's newly constructed kayak is lashed to the sled pulled by his Wainwright dog team. (Photo by Richard Nelson)

He swept out his little house and hauled duffel bags of gear to the beach. When the plane failed to show, he hauled everything back up the bluff. After several rescheduled flights also never came, Richard decided a string of dogs fueled by a pile of seal meat was the more reliable way to travel.

David Bodfish was also interested in traveling the ninety miles to Utqiagvik, so the two men hitched their dogs (five of Richard's and six from David's team) to the same sled and left the village in the early hours of April 23, 1965.

It was –6 degrees the morning they left, cold for late April. Richard sat on the sled, and David stood on the runners. Unlike the line of spectators gathered to watch Richard's arrival, his departure drew little attention. There was, after all, nothing unusual about two men and a string of dogs leaving town. David lifted the snow hook, and the team, eager and fresh, took off at a brisk pace. The village soon disappeared behind a slight rise as the dogs settled into a steady trot.

Perched on the sledload of gear, Richard watched small groups of caribou milling and grazing on the tundra. When he'd first glimpsed caribou, eight months earlier, they seemed strange, exotic, running beneath the plane's roar. Now he felt a comfortable intimacy with the graceful animals. He could, at a glance, distinguish the larger-bodied bulls from the smaller cows. He could recall the welcome heat as his frost-nipped fingers reached into a freshly killed caribou on a frigid day, steam rising tight and thick from its belly. The taste of heart tickled his tongue as he recalled steamy meals with the other hunters, smiles and laughter filling the tent. He carried the knowledge of how to skin the lower legs to preserve the hide for the making of tuttuliks, which he now wore on his own feet. As he bounced over the snow and ice, he appreciated the bulky fur parka that kept his body warm on this cold morning.

Months before, in the fall, Richard had told his parents how, over time, the vastness of the Arctic landscape fades from awareness.

> *It's such a great big monstrous area that you don't even think of it as wilderness. It just doesn't look wild. I think most people would be appalled by it. It takes a special person to be moved by this place. In its own flat, desolate way, it's really beautiful.*

The once-unremarkable tundra was now textured by stories, names, memory, and movement. As the sled slid east, Richard saw a pair of snowy owls hunkered on the banks of the Sinaruruk River, where he'd trapped foxes with Taqalaq. A few miles later, the sled tipped down the bank of the Kugrua River, followed by the Itinik and Tuvak Rivers.

Richard and David were constantly on the move, stopping only once for a snack of raw caribou and hot tea. They crossed hundreds of fox tracks, most going or coming from the sea, others trotting parallel to the coast. Tiny trails of lemmings skittered to and from small round holes in the snow.

David knew of an abandoned cabin on the north side of Peard Bay, which they reached by midevening. After staking and feeding the dogs, the two men spent an hour shoveling drifted snow and chipping ice from the floor of the rough shelter before they could rest for the night.

They arose early, fed the dogs in the dark, and broke camp just as the sky began to glow. When the sun rose, Richard turned on the sled, facing south to funnel the meager warmth into his parka hood.

As the dogs pulled him toward home, Richard didn't know memories of the teasing, smiling hunters would drift through his mind every day of his life. He could not fathom all the dogs he would come to own, all the trails he would come to travel. He could not imagine himself as an old man, an elder in his own community, awakening from dreams peppered with Iñupiaq words.

PART II

Making Prayers

KK'ADONTS'IDNEE

In decades past, Koyukon Indians lit their homes with wicks pressed into a shallow bowl of bear fat. The lamp required constant care—someone pushing the thick grease to the flame's edge so it would melt and burn. Koyukon people also burned long wands of splintered wood one after the other. But bear fat was hard to come by and the wood lanterns filled the house with smoke, so many of the long winter evenings passed in inky darkness.

Catherine Attla, born in 1927 on the banks of the Huslia River in Alaska's interior, grew up in those dark cabins. She remembers lying on a mattress filled with moose hair and ptarmigan feathers as her grandfather told stories from Kk'adonts'idnee, which translates as "In Distant Time, it is said." The prodigious number of stories from the Distant Time range from the minute to the cosmological; they describe a primordial world in which animals had a human look and spoke a human language (Koyukon). When these human-animals transformed into modern forms, they imbued north-woods creatures with human qualities and personalities. Koyukon people may inquire about a person by asking, "What animal is he?" A treacherous person might be described as "just like a sucker fish." A boastful person who promises a lot but accomplishes little is said to be "just like a raven."

Curled up beneath furs, Catherine learned those far-off ancestors metamorphosed not only into modern animals but also into the sun and moon, thunderstorms and mountains, birch trees and rivers. Everything, her grandfather Olin said, is alive and related. Everything has a spirit. Everything is watching.

It was taboo to tell the long narratives in the lengthening days of spring; they were reserved for late fall and early winter. At the end of a story (some took a month of evenings to tell), Grandfather Olin would say, "I thought winter had just begun, but now I have chewed off part of it."

In addition to the stories, there were lessons about *hutłaanee*, the elaborate taboos that govern Koyukon life. While skinning a beaver, Catherine's grandmother Eliza warned: "Don't eat beaver feet, or you'll grow up and walk funny." Out on the trapline, her grandfather admonished: "Don't say the mountain's name—your mouth is too small for such a big thing."

Catherine grew up in Cutoff, a village, now long abandoned, on the banks of the Koyukuk River. In the 1930s, when she was a small child, the village had no post office, no school, no church. The one trading post in town bought furs and sold what limited supplies the barge carried upriver in summer. "Had last candy around Christmas until June," Catherine later recalled. "That's when boat come. Eggs we had with the last boat in the fall. Potatoes same way. Onions too. We went without those things all winter long. That's when I learned a lot about how they used to do things long time ago when they were out of things at the trader post."

The art of listening and remembering formed the core of Catherine's education. She grew up without books or pencils and didn't begin speaking English until she was fourteen. She started teaching herself to read by studying soup can labels.

In her early twenties, Catherine lived on her own, catching whitefish with hand-tied seines and snaring rabbits with caribou-sinew nooses. Catherine said she felt "lucky there was no welfare back then," because it forced her to live close to the land, to take care of herself.

In 1944, Catherine married Steven Attla "because," she said, "it was too tough otherwise." Like Catherine, Steven's childhood was steeped in the Kk'adonts'idnee stories. Steven was a skilled hunter, trapper, and craftsman, who could split a tree into planks with wooden wedges and bend birch boards into canoe ribs. The young couple built a cabin in Cutoff and continued a subsistence lifestyle while raising nine children. When a school was established in 1950, Catherine attended classes two nights a week. She was thirty when she learned to read and write.

Richard met Catherine through his deepening friendship with Ray and Barbara Bane. In the winter of 1967–68, the Banes taught in the

Huslia school. Come spring, Barbara went to West Virginia to visit family. Ray stayed in Alaska and needed a buddy. Richard had two months free before beginning his doctoral work at the University of California in Santa Barbara.

It was a tense time in the United States: Martin Luther King Jr. killed in April, Bobby Kennedy assassinated in June, a steady line of caskets returning from Vietnam, student protesters met with tear gas and clubs. To Richard, the Koyukuk River sounded like just the place to be.

Richard arrived in Fairbanks on a late evening in mid-June. He unrolled his sleeping bag in a clump of spruce trees alongside the runway, avoiding the town's dusty roads and dingy bars.

The mosquitoes woke him with plenty of time to catch the early morning mail plane to Huslia. As they lifted off, low-angle sun bathed the sparsely forested hills surrounding the city. To the south, Denali, rising from the ragged stretch of the Alaska Range, cut into the brightening sky.

Hours later, the little plane descended toward an irregular cluster of forty or so houses dwarfed by an immense sweep of forested land. The pilot circled the village, then swung wide over a slough, leveled out, and rumbled onto a narrow sandy strip.

Huslia sits atop a high sand bluff on an outside bend of the Koyukuk River. In 1968, neat footpaths connected homes nestled among lovely stands of birch. Each place had a smokehouse and a meat cache perched atop tall spruce poles. Most homes had a team of dogs staked in the yard and a snow machine parked in a shed.

It was high spring in Huslia. Birch trees bristled with bright, fresh leaves. Juncos trilled from the tops of spruce. Snipes winnowed over a marsh. Robin song echoed across the river. Sled-dog howls lifted through town. These serene voices were a welcome relief from the nation's political strain.

Each evening, Richard and Ray joined villagers gathered on the riverbank overlooking the slow brown current and endless stretch of forest beyond. Near solstice, sunset bled into sunrise with liquid light. It was an easy, peaceful time. The young folks enjoyed fresh company; the older ones welcomed this inquisitive new white guy.

Richard's days were unburdened by deadlines and deliverables. He had no research project, no monthly report to write. Yet he couldn't help but ask questions, couldn't help being intrigued by Catherine, the friendly middle-aged woman who joined her neighbors each evening by the river.

Steven Attla was then a riverboat pilot, delivering supplies throughout the Yukon basin. He was away from home most of the summer, so Catherine came to the river alone. She was a small woman. Black hair in tight curls, round-frame glasses perched on high cheekbones. She welcomed Richard with a smile and answered his short questions with long stories.

After a few weeks, Ray and Richard, itching to get deeper into the wilds, hopped the mail plane to Kobuk Village. They assembled their folding kayaks on the banks of the Kobuk River and loaded in fishing poles and a month's worth of dried food.

Originating in the snowfields of the Brooks Range, the Kobuk flows west through a broad valley flanked by the Baird and Schwatka Mountains to the north, and the Zane Hills and Sheklukshuk Range to the south. For interior people, rivers are lifelines—sources of food and corridors of travels. The Koyukon language makes no reference to the cardinal directions. People orient themselves to the waterways: upstream or downstream, away from the river or toward the river.

Richard and Ray had no map. No place they needed to be. They camped on gravel bars, cooked on driftwood fires, awoke to the whispering river. An occasional rain shower, wind gust, or birdsong became the daily news; hordes of biting bugs, their biggest problem. As Ray recalled, "The wild beauty of the land, encounters with wildlife, and the constant anticipation of discovering some new wonder left little room for abstract politics. We were truly Huck and Tom caught up in our own world of adventure and discovery."

Their trip ended where the Kobuk met the Bering Sea. From Kotzebue, they flew back to Huslia and rejoined their friends on the riverbank. In late July, they loaded into Ray's newly purchased Citabria two-seat plane and lifted off from Huslia's sandy runway.

Ray flew north over an expanse of boreal forest, homeland to around two thousand Koyukon Indians living in a dozen villages along the banks of the Koyukuk and Yukon Rivers. The Koyukon belong to a cultural family collectively referred to as Northern Athabaskans. The Athabaskan homeland stretches two thousand miles from western Alaska to Hudson Bay and a thousand miles from its northern extreme in the Northwest Territories southward into lower British Columbia.

Ray navigated toward the rough-cut peaks of the Brooks Range, stopping for fuel in the tiny community of Bettles Field. They followed the John River Valley into the heart of the mountains. The maze of walls grew tighter and taller with each passing mile, part of the stunning landscape Richard described:

On this terrain the Athapaskan past has been played out and the array of cultures have been shaped to their present form. But very little remains here as a tracing of this human passage, so slightly has the land been altered. There are no visible scars, no straight lines, no crumbled ruins.

After refueling in Anaktuvuk Pass, they followed north-flowing rivers, touching down on river bars for the night. Evenings turned cold. The sparse tundra vegetation tinged red and gold. Richard's looming trip to Santa Barbara made the peace emanating from each campfire all the more precious. Soon enough he'd be stuck on California's freeways. Among rivers encased in concrete. Birdsong drowned beneath traffic whine. The politics of war polluting each day.

On Richard's last evening in Huslia, he joined Catherine on the riverbank. Low sun glinted off the water. A snipe winnowed in the distance. Mosquitoes buzzed nearby. Catherine asked Richard what he'd seen in his travels. He told her of a pair of loons on a clear Arctic lake. He described their brilliant white necklaces and checkered backs. With rising passion, he recalled their wailing call, one answering the other, the sound reverberating across still water.

Catherine nodded and smiled. "That's right," she said. "That's the spirit of that bird. You heard it. That's good."

They sat in silence, watching the river slide past.

Before turning in, Richard thanked Catherine for her stories. She smiled. "I'll teach what little I know to anyone willing to listen."

ICY COCOON

In the nine years after leaving Wainwright, Richard earned two graduate degrees, published three books, and walked away from four jobs. He was a rising star in the world of anthropology—book reviews in the *New York Times*, professorships offered across the country. In classrooms, he was a natural teacher, students captivated by his humor and passion. But lecture halls made him itch, faculty meetings made him fidget.

City living had no edge: Wait in line at the grocery store, and you were fed; obey traffic signals, and you were safe. Missing was the challenge of gathering one's own food. Gone was the vigilance required to survive on the ice. Lecturing about life in wild country was a dull substitute for the life itself.

Richard met Kathy Mautner while teaching in Honolulu. An aspiring anthropologist, Kathy shared Richard's fascination with other cultures. She was fit, muscles toned by years of surfing. She ducked beneath waves with ease, her long brown hair twisted into a thick braid. It didn't take long before Richard bought a board and joined Kathy in the lineup. Mornings, before classes, Richard and Kathy rode breakers thundering onto Sunset Beach. Evenings, Richard talked of trading their car for a team of dogs, dreamed of replacing highway hum with a whispering river.

In *Hunters of the Northern Ice*, he wrote that much of Inuit hunting culture would be "lost forever in the icy graves of the old men." His prediction seemed inevitable in the light of a towering yellow flame that flared into the Arctic night in the spring of 1969. The oil workers who ignited the natural gas roaring from the Prudhoe Bay test well stepped back to marvel at night turned to day. They'd discovered the biggest oil field in North

America. Eight thousand feet down, the pool of crude spans almost a quarter million acres and would, in the coming decades, become "worth more in real dollars than everything that has been dug out, cut down, caught, or killed in Alaska since the beginning of time," according to Alaska historian Terrence Cole.

For years, the flame burned unabated as the country's oilmen and politicians squared to the challenge of moving the black gold to market. Snaking eight hundred miles of forty-eight-inch pipe through three mountain ranges and over hundreds of rivers required bold engineering. Securing legal access across a patchwork of state, federal, private, and Native-owned lands demanded political solutions to lingering, contentious disputes.

Alaska's legislature, newly elected in 1959, eagerly took title to over a million acres of "vacant" federal land. Alaska Natives using and living on the land claimed the ground as their own. The Alaska Native Claims Settlement Act (ANCSA), signed by President Richard Nixon on December 18, 1971, was intended to resolve the land claims issues and pave the pipeline's path. The act granted forty-four million acres to twelve newly created Native-owned economic development corporations, each associated with a specific region of the state. To placate the environmental community, section 17(d)(2) of the act directed the US secretary of the interior to withdraw up to eighty million acres for conservation purposes. The selected "d-2" lands became proposed national monuments, parks, and wildlife refuges scattered across the state.

Zorro Bradley, an anthropologist with the National Park Service, was charged with assembling a team to assess the impact of newly protected lands on subsistence activities. Richard was in Hawaii when Zorro's call came. "You'd make a lot of maps," he explained. "We need to know where local people shoot ducks in the spring, harvest moose in fall. I need a team of people who can live on their own, travel and hunt with local people."

Richard knew his answer before he hung up the phone. Kathy needed a few days to think it over. When she agreed to move north, Richard called Zorro.

"We're in."

Zorro also hired Ray and Barbara Bane. After years in schools, Ray had traded his teaching certificate for an anthropology degree. The Banes, already living in the village of Shungnak, would help Richard and Kathy settle into the nearby village of Ambler.

In late December 1974, Richard and Kathy rode one last wave, stashed their surfboards, and left Hawaii for the Banes' house.

The plunge into winter shocked their beach-browned bodies.

> *Sixty-five below zero. The cold feels solid as if the air has become a super chilled liquid. It presses against you like water against a diver. It pinches your cheeks and forehead. When you breathe, the icy liquid fills your insides. The suddenly tangible air is frightening, an enemy, like cold fire. Walking up the Shungnak hill I was afraid to inhale the hard, burning air; but I had no choice. I tried to control the fear, lest it make me breathe harder—like a surfer stifling panic as he dives under a big wave.*

The cold snap persisted for days, dropping to –71. When it finally eased, in January, Ray and Barbara hitched up their dogs and took Richard and Kathy to their new home in Ambler.

> *The 35-mile trail is mostly over open country lying between the mountain slopes and the forested river valley. It is incredibly beautiful with the broad stretches of tundra, mountains pink with the low sun, and scattered patches of willow or timber. Temperatures around –10 degrees favored us and made the trip pleasant in every respect.*

Ambler, a cluster of cabins on the north shore of the Kobuk River, was home to ninety Eskimos and a handful of white pioneers. Mountains clung tight to town in the north and east. To the south and west, swatches of forest mingled with strips of open tundra, creating the most beautiful setting of an interior settlement Richard had ever seen.

As Kathy settled into Ambler, her tan faded, but her enthusiasm grew. In a February letter to Richard's folks, she wrote:

> *We're really loving the lifestyle up here. We've got four dogs now and would like to get one more. Kipnuk is the female leader—she's getting better all the time. Jasper is her son, a young male with a gentle personality and sad eyes. I have to be careful of falling in love with*

him—he'd be a good pet. Then we have Ray's dog Dakli, a huge male who is unbelievably strong and a bad fighter. And Tranjik, another terrible fighter.

Richard organized a public meeting to clarify his purpose in the village. With Art Douglas, a local man, acting as translator, he explained the information he gathered would determine boundaries and permitted uses inside the area that would eventually become Kobuk Valley National Park.

In the following weeks, Ambler residents shared coffee and pointed at maps to show where they shot geese in spring, killed caribou in fall, and trapped beaver in winter. Frank Jones, a hunter in his thirties, told Richard: "When I have the land and when I'm living in this country, I'm a rich man. I can always go out there and make my living, no matter what happens. Everything I need—my boat, house, clothing—is right out there. We must have this land, not just part of it but all of it if we are going to be sure of our living."

Old Man Cleveland, his wrinkled face creased into a smile, told him: "Eskimos should make laws for those people Outside. That would be just the same as what they try to do to us. We know nothing about how they live and they know nothing about how we live."

As Ambler folks shared stories from the past, the radio bore news of the future. In a letter home in March, Richard reflected:

I guess we're mighty lucky to be able to have all these outlandish experiences to live in unspoiled wilderness the way we do. It won't be long before all of this life has vanished and all we'll have is our stories. I often think about our being able to experience this wilderness by dog team—just the way people saw the early west from horseback.

As a kid, gripped with grief over the decimation of the Great Plains, Richard never dreamed he'd be involved in Alaska land conservation. The eighty million acres set aside by the ANCSA would, in the coming years, set the stage for the largest act of land protection in US history.

On December 2, 1980, President Jimmy Carter signed the Alaska National Interest Lands Conservation Act, which protected over one hundred million acres from industrial development. Richard's enthusiasm for the sweeping protections was tempered by his nuanced knowledge that wilderness was a needed balance to rampant growth but a blunt tool blind to an ancient way of life. In his journal, he explained the impact of the new law on the people he'd met:

> *When a man traveled, he was turned aside by no barriers except a natural one. When a man camped, he stayed wherever he pleased and for as long as need required. And when a man hunted, he took whatever he needed without regard for limit or quota set by anyone outside himself. This was the freedom he inherited, the basic right upon which all his activities were based as he moved over the land.*
>
> *This freedom has not been abused by Kobuk River people, who have taken according to their needs but whose needs have not exceeded the land's ability to provide. Proof that this system works over the run of centuries lies before us in the Kobuk Valley, where the land flourishes and freedom has never yet been curtailed.*
>
> *And if this system of interrelationships between men and the land has succeeded over the run of generations, is there need to change it today?*
>
> *The land here belongs to the Eskimos, all of it. They have had free use and access to this land, in effect ownership of it, for many centuries. The freedom they have known on this land is abruptly usurped, the land is claimed by forces outside of them and they are to be told how to behave on it. Can they feel any emotion other than fear or panic in face of this loss?*

In early April 1975, Richard and Kathy said goodbye to their Ambler neighbors and hitched the dogs. They mushed up the Kobuk River and met up with Ray and Barbara in Shungnak. The two couples planned to move their dogs to the Koyukuk drainage before the drip of spring turned the snow to slush.

Blizzard conditions mushing dogs over Dakli Pass (Photo by Richard Nelson)

After three days traveling beneath clear, calm skies, they awoke to low clouds and an uneasy breeze. The storm grew as they ascended Dakli Pass, the wind-blasted divide between the Kobuk drainage and the Koyukuk. Cresting the pass, ribbons of blown snow snaked past the sled. But the dogs trotted easily on the packed crust, their tails wrapped around their butts by the gale. Richard stood on the runners, wind buffeting his back. Kathy huddled beneath blankets in the sled's basket, her head ensconced in a furred hood. Richard leaned close, shouting through the squall.

"Warm enough?"

Kathy raised a mittened hand with a thumbs-up.

That evening, a few miles below the pass, the two couples staked and fed the dogs, then huddled against the storm.

"The tents will never hold up against this," Ray yelled against the wind's whine.

“Snow caves,” Richard suggested. “And a windbreak for the kitchen.”

Kathy and Barb freed a handsaw from the sled and cut blocks of hard-packed snow. The men unpacked shovels and dug into the face of a drift. Shelters finished, the men joined Kathy and Barb in the lee of their thick new wall and slurped a warm meal before turning in. Cozy in their sleeping bags, Richard and Kathy lit a candle in their tiny cave. The flame burned true, untouched by the crazed world churning beyond their icy cocoon.

FOREST OF EYES

Come morning, they crawled from their snow caves into a still, windswept world. Rounded lumps of snow shook, morphing into dogs eager for breakfast. The couples dug out the sleds, cooked a quick meal, and made the long downslope run toward the Koyukuk River.

In the years since first meeting Catherine Attla on the banks of the Koyukuk, Richard had returned to the village as often as life allowed. He had been captivated by the friendly people, the wild country, and the wise woman willing to share her stories. Each visit, he'd heard Catherine and other elders speak of a world where the creatures, plants, and clouds observed and listened.

The notion of a watchful world eddied and swirled through Richard's mind as he made his way to Huslia once again. In the Arctic, he'd learned to hunt seals and move over shifting ice. Could he now learn to move through a forest of eyes? Richard was unsure. But standing on the sled runners, dogs pulling toward the Koyukuk, he was excited to find out.

In Huslia, they arranged for someone to care for the dogs, then headed to Juneau for the summer. Richard hunkered down to write up the Kobuk work as coauthor of a book-length report: *Kuuvangmiut Subsistence: Traditional Eskimo Life in the Latter Twentieth Century*. In between writing stints, he dreamed of the coming winter.

Richard and Kathy returned to Huslia in early October, the tundra burnished red, the birch and aspens brilliant yellow. Steven Attla met their plane with an old red truck, one of three vehicles in the village. They tossed gear in the rusty bed and rattled in low gear along the town's single two-track road. The Attlas had offered Richard and Kathy the vacant log cabin tucked behind their own home. It was a tiny place, eleven feet by twelve feet, with a rough-sawn wood floor, no running water. A double bed was tacked along one wall, a counter tucked beneath a small window, a woodstove in one corner.

After stashing their gear, Steven suggested a sunset drive. All four packed into the truck's cab and bumped along a few miles of sand road. They stopped at the end, overlooking marshy, lake-filled flats. Steven kindled a small fire. They gathered around the flames, seated on downed logs. The sky blazed orange over the Purcell Mountains. No sounds, save the gentle snap of burning sticks.

Richard squinted through the smoke, watching Catherine poke at the flames. Wisps of gray now circled through her black curls. Richard looked over the blood-red sky. The expanse of the coming winter in this peaceful place with these gentle people felt like a precious gift. He brushed away a tear, turned back to the fire.

The next day, Steven helped Richard bring in a load of firewood and a bundle of logs to construct doghouses and a meat cache. Richard pounded stakes in the dog yard, which, he wrote in his journal, were spaced for "easy hitching after snow comes and setting the crazy fighters (Tranjik and Dakli) well apart." There were two additions to the team: Seelik, "a magnificent dog, like a sculpture, and very mellow and submissive," and Shungnak, just a puppy, the offspring of the Banes' prized lead dog.

On October 7—the year's first snow—Richard harnessed Shungnak to a sledload of dry grass for the doghouses. She was obviously an eager worker but too young for him to see if she had the required intelligence to become a lead dog. A week later, he and Kathy took Shungnak and Dakli out for a run. "Kath on the sled and me running along sputtering and shouting all sorts of instructions," he wrote. "Despite my antics Shungnak did fine again."

A short time later, Steven invited Richard to "go look for grouse." They rode double on Steven's snow machine, following a sand ridge running northeast from town. The skies cleared as they traveled, magnifying distant snow-covered mountains in the crystal air. As they traversed vast, brushy flats, Richard stretched his neck upward in an effort to see around. The terrain had a sunken feel, reminiscent of being in the trough of a huge ocean swell.

Richard and Kathy's Huslia cabin (Photo by Richard Nelson)

Throughout the afternoon, Steven navigated to over a dozen bear dens dug into subtle hills along creeks and meadows. Richard quickly realized "look for grouse" was a euphemism used in humble deference to the bear's spirit.

Each den had a name and story, which Steven shared as he looked for freshly broken twigs and newly dug dirt. He tossed handfuls of snow into the dens to highlight the shape and contour of the entrances. He also stomped about, knowing the bears often growled when disturbed. Although no animals emerged, Richard's admiration for Steven swelled throughout the day.

In the coming months, Richard would accompany Steven and other hunters in the search for "grouse." He'd watch successful hunters slice the eyes of a harvested bear to keep the animal from seeing any potential violations of the community's elaborate taboos. He'd learn how eating the brains would cause a man to anger easily, and how a woman eating any part of the head risked spreading rage through multiple generations. Standing with other men, visceral fat sizzling over a backwoods fire, Richard would come to see bear meat as communion, the mixing of humanity with the potent swirl of the land's spirit.

On the way home from the first bear-hunting trip, Steven stopped near a patch of water, kept open by a current bubbling in the bend of a stream. Four otters surfaced and frolicked in the snow. The scene enthralled Richard. Steven smiled, happy to see his new neighbor watching the world.

They returned to Huslia near dark and joined Kathy and Catherine for a meal of caribou stew. Near midnight, Richard and Kathy said good night and walked through a stand of birch to their own cabin. Kathy crawled into bed. Richard banked the stove, sat at their tiny table, and scratched notes into his journal:

> *These people are obviously bear hunters of the highest order. I was amazed by Steven's skill at everything he did today. Obviously, he is a masterful outdoorsman.*
>
> *After dinner Catherine and Steven talked about someone recently catching a live junco (bird) and caged it for a pet. Asked if this was not hutłaanee, they replied that an old man at Hughes caged a bird for 6 months with no ill effects—so they concluded that this was alright.*
>
> *Obviously, hutłaanee taboos are viewed with some flexibility. They are not considered inviolate, in many cases, until someone has tested them. Young people or outsiders may be able to break a taboo without ill-effect, but this does not mean that adult residents can do so. The best test is to have your peer experiment, and the result can be your guide.*
>
> *One of the oldest village women had followed the taboos faithfully all her life, and she is still completely alert and strong. Two others have not, and their health is far poorer, even though they are not as old.*

Later that month, the temperatures plunged from 0 to –34. Richard hitched the dogs in the biting cold and set off to retrieve a pile of firewood. Shungnak, frisky yet obedient in the lead, kept the team from floundering off the packed trail. At the woodyard, Richard set the snow hook and hauled green logs through thigh-deep snow. After a few dozen trips, Richard paused to calm his racing heart. Sweat tickled his ribs. Cold stung his cheeks. The low midday sun sparkled in the snow-laden trees.

Back home that evening, the cabin walls creaked and popped, caught between the stove's radiating warmth and the night's penetrating cold. A dog barked—once, again—then lifted into a full howl. One by one, the rest of the team joined in. Exhausted from the day, Richard forced himself to journal before joining Kathy in bed. He described the hard work of moving logs in deep snow.

> *But being outside in this magnificent wild country is more than ample compensation. We are lucky people, and we don't forget that.*

They awoke to more clear and cold. Richard again hitched the team.

> *I went to cut dry willows along Racetrack Slough. It was late afternoon, and the sun was swollen and red near the horizon. Along the slough the snow reflected red, and the dogs' breath was red, and flakes of frost scuffled by their feet glowed red. It was a fine trail, so the dogs galloped happily, Shungnak almost dancing along at lead. Ahead of us a fringe of tall thin spruce silhouetted into the sunset. While I cut willows, I saw the sun enlarge and flame red, flattening itself into the spruce at the horizon's edge. Is there any wonder that we love the way we live?*

LUCK

Chief Henry was the oldest man in Huslia. He lived with his wife, Bessie, near the Attlas'. The first time Richard knocked on their cabin door, the old man waved him in. He wore black suspenders over a plaid shirt draped over his thin shoulders. His eyes, milky white with cataracts, peered out beneath unruly gray eyebrows. When asked his age, Chief Henry said, "Somewhere close to ninety, I think. Not too sure."

Richard sat beside Chief Henry on a bed that doubled as a couch in the simple log home. Bessie, a youngster of eighty, served tea. Chief Henry and Bessie spoke softly, with long pauses between their stories. They shared memories about visiting Yukon River villages long ago, by boat and dog team and recently again by plane. Bessie then lifted a framed wedding license from the wall, dated 1910. It was a prearranged marriage, she told Richard: "We only talk one, maybe two times before we got married. We got real lucky. Sixty-five years now. We've been happy the whole time."

Over a second cup of tea, Chief Henry and Bessie talked of the early trading posts and the first time they were handed a can and told it was food. "We just look at that thing," Chief Henry said, "and thought, 'How the heck we supposed to eat this?'"

"Nowadays," Bessie said, "we are lucky. Moose and caribou live around here."

"Used to be," Chief Henry added, "all we had was fish, ptarmigan, and rabbits. If it wasn't for those rabbits, we wouldn't be alive today."

The old couple related a particularly rough spring. The lakes had begun to thaw and the first ducks had arrived when a cold snap gripped the region. Open water refroze, sending the waterfowl south again and locking muskrat and beaver beneath the ice. Hares were difficult to snare, because crusted snow freed them from established trails. Bigger game could easily hear the crunch of an approaching hunter.

"We got so weak," Chief Henry said, "we could do nothing but lie down. If one more day passed, then I would have to go to somebody's camp and ask for food. But that day I got lucky—caught a rabbit and a marten—lots of grub then. So we stayed on the trapline."

"We prayed a lot that spring," Bessie said. "We showed our respect for the animals. All we had was our luck."

Most every page of Richard's Huslia Journal explores the notion of luck. And Richard came to understand luck as "a tangible essence, an aura or condition" holding sway over every aspect of life, which he later explained this way:

> *People who lose their luck have clearly been punished by an offended spirit; people who possess luck are the beneficiaries of some force that creates it. Koyukon people express luck in the hunt by saying bik'uhnaatłonh—literally, "he had been taken care of."*

Everyone is born, Richard was told, with a certain amount of luck. The difficulty is not so much in getting luck as keeping it. In Wainwright, his journal had filled with stories of people hurt or killed by sudden shifts in sea ice. In Huslia, stories centered on sudden shifts in luck.

A man who spoke of his plans to trap many beaver didn't catch a single one all season. Another man who boasted about his bear-hunting skills was later attacked and hurt. Bragging about hunting an animal is hutłaanee because it shows disrespect, "like pointing or staring at a stranger." Their change in luck was attributed to their forbidden words.

In addition to respectful language, good luck required proper action. When skinning an animal, the animal's name must not be mentioned and pungent smells and metallic noises are avoided. Some people take the added precaution of wrapping an animal's head with a cloth and filling its nostrils with lard, to prevent any chance of offending the animal's spirit.

The bones of water animals (muskrat, mink, and beaver) are cast into a lake or river. The bones of large land animals are returned to a dry place, well away from the village. The remains of small animals are hung in bushes or completely burned.

Only old people who no longer hunt can eat red-necked grebe, because the bird is awkward on land. Young people who consume the bird will become slow and clumsy. Pregnant women must not eat beaver meat, lest their children come to walk with inturned feet.

Objects, too, are imbued with luck. The first use of a new pair of snowshoes or a new sled should always be in a downstream direction, because this is the direction spirits go. Rifles are said to become "luckless" to the point of uselessness. Putting on another's mittens can take a person's luck or give yours away.

Trees and mountains have an awareness sensitive to offensive behavior—as does the weather. If a man brags that a storm or extreme cold could not stop him from doing something, it's said that "the weather will take care of him good."

Steven Attla mused to Richard about a luckless year trapping foxes. "I was using power tools while I had a fox in the house. Guess that's why . . . It's got really sensitive ears," he said. "When you get bad luck like this, you just have to let it wear off. There's nothing else you can do."

Richard came to see keeping luck as a state of grace—less about fortune, more about proper relationship with each animal and plant, every mountain, lake, or river slough, and all the shifting moods of weather.

Like smoke making wind visible, luck brought shape to the forces hovering just beyond the edge of sight. And there were a thousand ways to lose it. Once gone, no amount of skill could replace it. "In the absence of luck," he wrote, "there is no destiny except failure."

All fall, as Richard and Kathy listened to stories from the Distant Time, an army of workers fused sections of pipe, pointing toward Alaska's future. A few hundred miles east of Huslia, welders, crane operators, and engineers labored in the frigid cold beneath the bright glare of outdoor lights. In Fairbanks, hub of the construction boom, local businesses struggled to compete with the high-priced jobs. Brothel business was brisk. The local McDonald's served more hamburgers than any other outlet in the world.

Risk rewarded with riches—this has always been the pioneer's promise, the drive behind every race for land and every rush for gold. The pipeline was a new day for an old dream—never mind that the "last frontier" had long been someone's home. On their days off, workers explored the endless wilderness. They zipped along rivers in shiny new boats and tested their marksmanship on the abundance of game.

Huslia hunters found moose carcasses along the banks of the Koyukuk River, missing only the head and a few choice cuts of meat. Such waste was an affront to the Koyukon belief that any creature should be carefully treated and fully used to avoid offending its protective spirit.

Increasing encounters with skinned bears and poorly butchered caribou stirred a growing animosity throughout the region. Town meetings swelled into rants against the influx of outside hunters. The older folks were concerned but calm. The younger folks called for militant resistance to all whites on the river.

When representatives from the Alaska Department of Fish and Game and the federal Bureau of Land Management (BLM) arrived in Huslia to discuss public easements across Native-owned lands, there was no ambiguity in the community's response. No outsiders would be welcome for any reason—no hiking, sightseeing, camping, or fishing. "Once they come in for anything," one resident said, "next thing they'll start killing our animals and spoiling the land."

When the BLM man reminded folks that the government had just deeded land to the community, a young man shouted, "You can't give us what we have always owned."

The bureaucrats fell silent.

Another man spoke sharply, "For us this land is a supermarket; for them it's a playground."

Richard slipped out the side door and made his way to the calm warmth of Chief Henry and Bessie's home. Bessie, as always, served tea. Chief Henry mused about the tension in town.

"Some of these young kids, they asked me if it would have been better if the white man never came around here in the first place. I just looked at those kids and said, 'Did you ever have to keep alive by eating ptarmigan droppings?'"

Chief Henry rubbed his whiskered cheek with a gnarled hand. He sipped his tea and, with a faraway look, said, "We went through some hard times all right, but I never saw anybody starve to death. This country always took care of us."

When Chief Henry began to nod off, Richard slipped into his coat, thanked Bessie for the tea, and said there would be plenty of time to talk later.

"Not for me," Chief Henry said. "Not much more time for me."

SKIPPING HEART

Kathy pinned a birch log to the sawhorse as Richard cut rounds. Both hands on the bow saw, he rocked back and forth with a steady rhythm. Each push and pull spit wood dust onto the snow. When a round dropped, Kathy nudged the log forward and Richard set the teeth, cut again. They worked without words, stepping through a well-rehearsed dance.

Warmed by the work, Richard shed his hat just as Catherine and Steven stepped through the trees. Catherine imitated the sound of steel teeth zipping through frozen wood. "I learn to make that noise so I can have that song in my mind whenever I want," she said. "The handsaw always makes me feel strong."

After helping stack rounds against the cabin, the Attlas invited Richard and Kathy on a trip to the graveyard. It was Thanksgiving. "The ancestors should always feast first," Catherine explained.

On a hillside scattered with fenced graves, Richard and Kathy watched their friends scrape snow from the base of a wooden cross. Catherine piled curled strips of birch bark on the frozen ground. Steven struck a match, sheltering it with a bare hand. Fingers of flames climbed through the dry bark. He added larger sticks. Smoke curled into the still air.

Catherine fed a chunk of meat and a piece of dried fish to the flames. They watched the meat sizzle and char before she spoke. "Grandmother, I just want to say that you told us so many things when you were living and you lived such a long and good life. These two people are good friends, and I wish you would help them to live a long life and have good luck."

They returned to town in the dim light of late afternoon. In the evening, the two couples shared a meal of roasted goose, potatoes, and wild cranberries. After blueberry pie, Richard stacked the dishes and joined Catherine at the kitchen sink. Catherine washed; Richard dried. "You take such good care of us," he said. "You feel like our parents."

Catherine laughed. "Me and Steven was thinking the same thing."

TEMPERATURES DROPPED TO −45 DEGREES. Frost crystals condensed and fell from the clear, superchilled air like a dense, glittering fog. Each night, the fire burned out and Richard and Kathy awoke to iced-over water buckets and coffee cups frozen to the counter. Each day demanded another trip to the woodyard to satisfy the hungry stove, but their spirits remained high.

Their contract with the National Park Service required Richard and Kathy to submit monthly reports answering where and when Koyukon people harvested game. But personal interests pushed well beyond professional obligations. Apprenticed to Catherine and other Huslia women, Kathy learned to tan hides, cut fish, and sew furs. And, caught in the current of stories, Richard explored new pathways of perception.

One night, dinner over, dishes done, Catherine spoke to Richard about her first encounters with Christianity. "I was twenty-three when we got the first church," she said. "I couldn't read the Bible so I just listened to the preacher's stories. Pretty soon I got all mixed up. Didn't know what to believe. My grandfather told stories to bring the people good luck, keep them healthy, and make a good life. Preacher said the same thing. When my grandfather came to songs in the stories, he sang them like hymns.

"That preacher got real interested in my grandfather's stories. When I told him I didn't know what to believe that preacher told me, 'You have to carry both stories. Both are right. People were given power from God to heal one another.' I sure felt good after he told me that."

Catherine echoed these thoughts in Huslia's small church, where the local pastor shared the pulpit with community elders. That Sunday morning, she shared her conviction that prayers would be heard whether they were spoken in English or Koyukon. "I've thought about this a long time," she told the congregation. "No one has to choose the white man's way or the Indian's way. Christianity works for all people everywhere on Earth, including us. But the Indian way works too. We should be proud and follow our own ways, while also following those that have come from the Outside and seem useful."

Chief Henry spoke next. With a quiet voice, he talked of the need for religion to flow through the generations. He gestured to the gray-haired church members and asked, "Where are the kids?"

December 1, Richard's thirty-fourth birthday, dawned clear and cold. The low winter sun was too weak to warm the nighttime low of –35 by more than a few degrees. In the early evening, Steven and Catherine stopped by for a quick cup of tea and stayed until midnight. As the cabin logs creaked in the deepening cold, Steven talked of the loud cracking sounds made by lakes as the ice thickens. "When this happens," he explained, "the lake ice is asking for snow to come and protect it from the cold. Even the ice has a life in it."

Catherine talked about how her grandfather Olin had healed her from multiple sicknesses. He never used plants or potions—just water and song. Often, Grandfather Olin invoked K'onghabidza, the raven spirit, to "scare away the sickness in someone." He'd mimic the raven's melodious cawing, spread his arms like wings, and hop on both feet like the bird. The old shaman, she said, had cured Steven too. Richard was struck by the sudden loss of traditional medicine.

> *Today there are no medicine men, the tradition is entirely lost. Steven and Catherine have seen and experienced it; their children will never see it, but can only be told. So close and yet so far. It could never be revived, and so we can watch it fade and be forever lost.*

Early Christmas morning, Richard's heart literally skipped a beat. Then another. Then another. Richard stared into the dark, Kathy asleep by his side. For more than three decades, Richard had never thought about his faithfully thumping heart. But miss a beat, and he could think of nothing else.

He woke Kathy.

Ear on his chest, she listened, eyes wide in shared fear.

In a city, Kathy would have warmed the car and driven Richard to the emergency room to be scrutinized by the blink and beep of a cardiologist's machines. In earlier times, they might have sent for Grandfather Olin and watched as the old shaman waved feathers and chanted to steady the rhythm of Richard's heart.

With limited options, they bundled up and knocked on the door of the village health aide. Still in her nightgown, she listened through her stethoscope, then hailed the nearest doctor (130 miles away) on the VHF radio. "Sounds like stress," said the doctor. "Relax. Take it easy. Merry Christmas."

Huslia was caught between worlds. Shamans gone, doctors not yet there. Animals respected by locals, poorly butchered by distant hunters. The old stories fading in the minds of elders, the new stories not yet formed. It's hard to relax when a life-sustaining rhythm stutters. Richard spent Christmas evening at home, waiting for the next missed beat.

By the evening of the twenty-sixth, things had improved—just a few skips each hour—but Richard vowed to make some changes.

> *I have to learn to worry less about the project, stop working all the time. With fieldwork you don't quit at 4:30 PM and go home to watch T.V. Twenty-four hours a day, it's easy to find work to do—pay a visit, take notes, have an interview, or any of the thousands of daily chores. It's hard to believe we can create pressure in such a peaceful setting, but we bring our Protestant Ethic to the village with us. Much as I despise the Protestant Ethic, I am inclined to become its victim. But the body has sounded a warning. I will slow down.*

In the New Year, temperatures dipped to –40 degrees. Kathy, having nearly frozen her feet during an earlier outing, decided to stay indoors when Richard proposed a run with the dogs. He ran the team north of town, threading through thick willows and spruce. The sun was just beginning to set as the team emerged from the trees and loped along the frozen expanse of an open slough. At the far end, Richard slowed Shungnak to a stop. He set the snow hook and looked back over the flat expanse.

Breath from the dogs had condensed into a snakelike cloud stretching for more than a mile over the snow. Iridescent in the twilight of sunset, the fog gently contorted into misty peaks and spirals. As the sun dropped, the sinuous cloud flamed orange against a background of dark trees. A single raven, like a black spot on a white page, flew over, the rhythmic rush of air through feathers loud in the silence. The bird tucked its wings, twisted into a brief free fall, called once, and disappeared between the tops of spruce.

That evening, Richard turned to his journal.

I remember, when I was a boy, walking alone into a huge, beautiful, darkened cathedral. My entire body was alive with a sensation of being watched—by the walls and windows, the pews and pulpit, by the air itself. Now I have felt it again, but this time when I was traveling alone in the forest. It is hard to imagine a more profound peace.

ORPHANED RAVENS

Cold gripped the back of Richard's throat as he stepped from the cabin. He glanced at the thermometer: –37 degrees. The rising sun cast a pink sheen over high gray clouds. Chains rattled as the dogs emerged from their houses in anticipation of breakfast. Richard tossed a scoop of food to each husky. After they'd gulped it down, Richard knelt to scratch Shungnak's ear. Shungnak caught his whiskered chin with a few quick licks.

A raven called from a nearby spruce. The bird dropped to the ground and strutted across the snow on the far side of the dog yard. The raven cocked its head and took a few tentative sideways steps toward a piece of dog food that had rolled beyond the reach of Shungnak's chain. The raven jumped back, just shy of the prize.

Again, the wary sidestepping approach. With a quick lunge, the bird snatched the brown kernel in its beak and twisted through the trees, wings huffing in the still air.

Only in the last two to three years had the birds foraged near people's homes. One accomplished dog musher in town told Richard about a talk he'd had with a fearless raven hopping between his dogs. "Go ahead—eat with the dogs," he'd said. "But then please make them pull well." The raven then took off and never returned, leaving the man anxious.

Catherine and Steven fretted about the change as well.

"They don't belong here," Catherine said. When Richard pressed her to explain, she added, "Since the medicine men disappeared, no one knows how to use the animal spirits, so now they drift around like orphans. These ravens nowadays feel like orphans."

She went on to share a story from a recent trip to Fairbanks. "I saw a raven sitting on a streetlight pole. It looked all oily and messed-up, just like it didn't take care of itself at all. I was upset about this, so I looked around to see if anybody could hear me. Then I just talked to it like my grandfather would. I said I wished it would go out to where it could live off scraps of hunted animals. Then it wouldn't look so poor and helpless."

ABOUT A THOUSAND MILES FROM Huslia, bureaucrats in Juneau were crafting the state's policies on subsistence. Increasing hunting pressure had forced lawmakers into a contentious debate over who got access to animals. Did sport hunters have the same rights as subsistence hunters? How could priority be given to one group over another? When Zorro Bradley, the closest thing Richard had to a boss, asked him to attend legislative hearings, Richard decided to do so.

> *I sense how easily people could lose their connection to the land and its resources unless the state is managed with subsistence in mind. I hope our work is of some benefit. It panics me to think of people like Steven and Catherine losing their beautiful lifestyle.*

While packing for the trip, Richard's heart once again began to stutter. In between meetings in Juneau, he slipped out to see a doctor, who confirmed the arrhythmia was stress-induced. The doc prescribed "hypnosis, meditation—whatever it takes for you to relax." Long days in airless rooms talking with suit-and-tie politicians was not what the doctor had in mind.

The first night back in Huslia, the cold calculations of distant politicians faded in the familiar warmth of the Attlas' kitchen. Over a meal of caribou stew, Catherine spoke of her grandfather's trapline. "It's a funny little place," she said. "It's got so many fox, beaver, muskrats, and mink. All around, there's not so many animals. But right where my grandfather trapped, there's lots."

She went on to explain the richness was no accident, that Grandfather Olin had used his power to attract game from surrounding lands. "He died almost twenty years ago, but his medicine has not worn off yet," she told Richard. "Someday it will, and the animals will leave his trapline."

Richard loved hearing cultural gems like this.

It's exciting to start learning again, like taking an excellent course at the university, but, in this case, the material is unwritten and from a world so different that it is beyond imagination. It's a privilege experiencing this lifeway before it dissolves into our own, to learn at least a part of this beautiful culture. I choose that word carefully—this culture is beautiful. Its people have a softness and sensitivity that rests well with me. They manage to combine the practicalities of subsistence with sensitivity toward the beauty of their nurturing environment. This balance of pragmatism and sensitivity is, to me, the ideal form of human adaptation.

Excitement for learning offered no escape from the winter's cold. For a week, the thermometer never crept above –40 degrees. Like a starving person dreaming about food, Richard and Kathy flipped through thumb-worn surfing magazines and reminisced about their favorite Hawaiian beaches. In the cramped Huslia cabin, they recalled the wide sands and solitude they'd enjoyed on a fall camping trip near the coastal community of Yakutat in southeast Alaska. They decided, come summer, they'd return to Yakutat—maybe buy a piece of ground there to call their own.

Eager for a change in routine, Richard and Kathy hopped the mail plane to Hughes to spend a week with Ray and Barbara. Although just forty-five minutes apart by air, the two couples had not seen each other for months. It was a treat to slip into the ease of each other's company, and Richard found the energy there to be just what he needed.

Hughes is a quiet, slow-paced, almost somnolent community. Such a contrast to the hyperactivity and intensity of Huslia. The atmosphere is relaxed and pleasant—people are easy to be around and contented with a minimum of change. The scenery contributes to the sense of peace. Hills and mountains are all around, giving a feeling of wildness to the place. Both Huslia and Hughes are extraordinary villages but for very different reasons. One is notable for beauty and tranquility, the other for vigor and strength of personality.

On Ray's recommendation, Richard paid a visit to an old man named Lavine Williams. Lavine lived with his wife, Suzy, in a slumped log home on the upstream edge of town. Richard knocked on the low wooden door and soon found himself cradling a cup of black tea on a tiny table. The panes of the cabin's only window drooped with age, blurring the view of the river. Lavine spoke English with a heavy, rhythmic accent typical of Koyukon Indians.

Richard explained his interest in animals of the boreal forest.

"What animal you want to know about?" Lavine asked.

"How about the kingfisher?"

"No. That bird is dirty. Lives in a hole. Sometimes when it catches a fish it leaves it to rot. I don't like that bird. Let's talk about the loon."

In Kk'adonts'idnee times, Lavine explained, the loon used its medicine to restore a man's sight. The man had a dentalium necklace, which the loon kept after making its medicine. And this is why the loon has that white pattern on its neck.

"If you walk up to a lake," Lavine said, "and a loon makes one call and dives, you gonna have bad luck. But if two loons run across the water together with their wings back like this"—he pulled his arms behind him, hands cocked toward the floor—"that means good luck."

"Long time ago," he continued, "when the old-timers want to make a new song they always go listen to the loon. They said, *'Ots'aa dodzin tong-heedo beeznee eeneetłon,'* which means 'When a loon calls on a lake, it's the greatest sound you can ever hear.'"

Richard returned each afternoon. Some days, Lavine seemed grumpy and complained about the young people zooming through the countryside on snow machines. "They go through the woods, but it's like they were never there. They live in a cloud of noise. Too fast. By the time they might see something, it's too late. And nowadays their mothers feed them cow's milk. No wonder they act like animals!" Lavine took a sip of tea and looked out at the river. "Too much education; not enough mind."

Other times, Lavine welcomed Richard with a mischievous grin. As they each gripped a tea mug one March day, hands almost touching in the center of Lavine's tiny table, the old man asked, "Who you want to know about today?"

"How about red squirrel?" Richard began.

"*Tsaghaldaala*, they call it. Means 'climbs a tree quickly.' In the old days, if you eat a squirrel you maybe starve later on. The skin is good for parkas or inside of mittens. Real warm and does not shed.

"It's hutłaanee to burn a squirrel. Better to leave it out in the woods for the others to eat. If you hear one running up and down a tree at night and talking loud, it's a bad sign. I only heard that one time. I was with another man, and that guy died not too long after that."

"How about sandhill crane?"

"They call it *dildoola, dildoola, dildoola*." Lavine repeated the word to illustrate how the name mimics the bird's call. "In the fall, you see the parents teaching the kids how to fly. That's when they teach them to talk too. There's no hutłaanee—they're real good to eat. But I don't believe in killing them myself. I like to just listen."

Every bird mentioned—yellow-shafted flicker, northern three-toed woodpecker, white-winged crossbill—Lavine knew its natural history and lore and could sing its songs.

Back at Ray's house, Richard stayed up late journaling.

> *Obviously, he's an animal watcher with an eye for the aesthetic qualities and personalities that each species manifests. It is so exciting for me to learn these new perceptions of nature; I just can't get enough of it.*

A few hours before his scheduled flight back to Huslia, Richard stopped by to thank Lavine for all he'd shared. They chatted awhile outside the squat riverside cabin—Richard bundled in a parka, the old man in a flannel shirt, impervious to the biting cold. As they shook hands, Lavine looked his new friend in the eye and said, "Just remember, every animal knows way more than you do."

RICHARD LAID A MAP ON the table. Steven and Catherine traced routes from lake to lake with their fingers, place names rolling off their tongues with increasing excitement. The complex words pushed Richard's linguistic abilities as he scrambled to keep up with a torrent of unfamiliar words.

Ts'aatiydinaadagondin (Huslia): "where the hill comes down to the river"
Notozahtłdin (a lake): "lake shaped like a hook"
Dodzin nohk'aa (a lake): "common loon lake"
Toonoonghun' (a lake): "blackfish with its stomach full of water"
Gohtłitłt'o teeya (a hill): "nape of rabbit's neck hill"
K'odzghoteel teeya (a hill): "where they hunt with bow and arrow"
K'idotł'on da-oy dakok'a (a lake): "fish trap funnel lake"

In his journal, Richard catalogued hundreds of the place names the Attlas told him about.

They have given 150 names in two evenings and obviously know many more. The density of names exceeds all expectations.

I'm getting lots of practice writing Koyukon. Surprisingly, this language is no more difficult to hear and write than Eskimo—I sure hadn't expected that! The only problem is the length of the names, which defies all imagination.

Looking over the growing list of names, Steven mentioned that he and a few other men had talked for years of doing this themselves. "If I die," he said, "then the country will die with me." When pushed to explain, he just shook his head and stared at the map.

Catherine then spoke. "The young people nowadays are not listening. They're not learning the stories. They don't know the names of the places. Maybe that's why we have too much bad luck nowadays—too many things go wrong. Because some younger people don't know how to treat things the right way."

Catherine paused and looked from the map to her husband, then continued. "We need the land to survive. But the land needs people too. People need to pay attention. That's what keeps the balance."

Decades later, thinking back to that map spread out before his dear friends and greatest teachers, Richard reflected: "I was witnessing the death of knowledge. And not just any knowledge. The rituals and sensitivities and stories of the Koyukon people comprise what I believe to be

Richard taking a day to get caught up with field notes in Huslia (Nelson archive photo)

humanity's greatest religious tradition," he said. "Their worldview holds the greatest potential for humanity's future on this planet. Without integrating some restraint and humility, there is no hope of a sane future.

"I realized that no matter how hard I worked," he continued, "I could only write down a tiny fraction of the rich knowledge in the minds of just two people. I felt impotent. Whatever detail I let slip by, any idea I failed to capture in my journal—it felt like it would forever vanish."

SPRING

Richard and Kathy staked the dogs in fading light. Using their snowshoes as shovels, they scraped through two feet of spring snow and piled a thick bed of spruce boughs on the exposed ground. They built a brisk fire as sunset deepened to twilight. Caribou meat sizzled over the flames. A pot of fish for the dogs bubbled at the fire's edge. A full moon rose from the slope of Bear Mountain and bathed the land in pale light and deep shadows.

Near midnight, −10 degrees, Richard tossed a few more sticks on the fire and joined Kathy in their zipped-together down bags. Peering from their feathered nest, they watched sparks wiggle into the moon-bright sky before dropping into dreams. The fire burned low, coals fading in the wide, cold silence.

Smoke from their breakfast fire scribbled across a cloudless sky. They were returning to Huslia after a rich week in Hughes. Richard had spent afternoons with Lavine. Kathy had worked with artisans, tanning and curing furs. Evenings, they relaxed in the welcoming warmth of Ray and Barbara's home.

Richard and Kathy hitched the team, the dogs barking and lunging against their harnesses. Richard lifted the snow hook; the dogs took off. Shungnak, confident in the lead, kept a brisk pace along the slick trail winding up Bear Mountain. They crested the pass, dogs panting, tongues hanging from the sides of their mouths. The downhill run was fast and wild. Richard rode the foot brake to keep the sled from overtaking the dogs.

As the terrain flattened, Richard noticed fresh caribou tracks moments before the dogs burst into a sprint. Ahead, three caribou ran down the trail, effortlessly outpacing the frantic dogs.

In the afternoon, the weather warmed to a balmy 15 degrees. Kathy pushed back her parka hood to soak in the day. Fox tracks meandered from grass clump to spruce sapling, each marked with a yellow stain. Furrowed paths of moose cut through patches of willows. Ptarmigan burst from the snow, startled by the dogs.

The sun was low but still visible when they arrived at an empty trapper's cabin situated on a low bluff overlooking Dulbi Slough. Dogs fed, Richard and Kathy salted strips of caribou meat hot from the fire. They lingered on the cabin's porch to watch the moon chase the sun from the sky.

Inside, candlelight glimmered off the ax-hewn logs and low ceiling festooned with dusty cobwebs. Kathy sewed while Richard wrote in his journal.

> *It's gloriously peaceful here this evening. We're tired but deeply satisfied traveling with our five dogs. Most people zoom over the country in an airplane nowadays, or by snow machine. The few who make the trip with dogs use 10–15 in their team and carry no load. We're traveling like old-timers.*
>
> *Kathy's sewing herself a new pair of beaver skin mittens, using hides she tanned herself. Looks like they'll be beautiful. Quite a lady, she is. Don't know how I ever got so lucky to find her. She's made the most of living here.*
>
> *I'd never let her go.*

It was midmorning before they began the easy three-hour run into Huslia. A few miles from the village, they met Steven out on the trail. They chatted in the sun, Steven curious about who they'd visited in Hughes, what they'd seen along the trail. Richard and Kathy hungry for the news of home: Had anyone seen geese? Had the river ice moved?

In the warming days of April, Steven and Catherine busied themselves outdoors, where they could scan the sky for geese. Evenings filled with stories of spring hunts and summer fish camps.

Then the soft snow hardened beneath a cold snap. And the chilly days spiked Richard's spring fever.

> *I find myself longing for slush and slop, for mudbanks and puddles—give me anything, so long as it brings an end to this*

eternal winter. I often wonder why we love Alaska so much, being people who rank winter dead last in our preference to seasons. I know I'll miss dog mushing but I'm hard pressed to find any value whatsoever in freezing temperatures, snow, lifeless land, bare trees, inert landscapes, and mounds of clothing. Every day seems to creep by as we await the end of these natural harassments.

It snowed five inches on the last day of April. By evening, the temperature dipped to zero. A stiff wind battered the cabin. Richard dreamed of birdsong—high, light melodies drifting through a snowy world.

In the morning, he rubbed his eyes and looked around the cabin, but the dream continued.

I heard robins!

When I got out of the sleeping bag the cabin was warm inside. When I stepped out, I couldn't believe the warmth—water dripped from the roof and the snow was wet in front of the cabin. I heard geese calling. I heard blackbirds. I heard sandhill cranes. Looking over the flats with binoculars, I could see flights of geese everywhere, flying north with the stiff wind behind them.

Richard harnessed the team. He mushed along lakes and sloughs alive with juncos and snipes, Bonaparte's gulls and arctic terns, Lapland longspurs and short-eared owls, whistling swans and hawks. Thousands of white-fronted geese streamed over, their high bugling call punctuated with the lower honks of Canada geese and the throaty trumpets of sandhill cranes.

How can this happen? It's like spring in the north is controlled by a switch. And what a glorious season it is, especially for those of us who exist in a state of suspended animation all winter. The air is warm and the sun blazes—we sat for hours outside doing nothing, like turtles on a log. We have paid dearly for this pleasure and will enjoy it to the fullest.

CHIEF HENRY'S SONG

The first Sunday of May, church service was held on the riverbank. Meltwater swelled the river and lifted the ice. It was a dangerous time, with ice-choked floodwaters capable of destroying homes. As massive floes cracked and ground together, Chief Henry thanked the ice for allowing people to travel on it all winter. The old man, one hand on his cane, gestured downstream and encouraged the ice to move on. "There's frog fat down there," he said. "Go on down, and it will be there for you."

For long hours each day, people gathered to watch the ice twist and turn on the thick brown water. "Current is too slow," Steven said. "There's a big jam upstream somewhere."

Catherine filled the time with childhood stories. "When the ice was running, they always kept us kids quiet. Just tell us, 'Shut up—your mouth is too small to talk about big things.' We weren't allowed to even throw sticks on it. That's how we learned about respect. That's how we learned about the spirit that lives in everything."

Looking over the ice-filled river, she recalled the year the military tried to break ice jams on the Yukon River. Low-flying jets, a string of bombs, towers of brown water exploding from the ice. Richard tried to imagine the pilots in those planes. Young men, raised in Topeka or Pittsburgh, maybe even Madison, who'd enlisted to serve. Smart and ambitious, rising through the ranks, awarded their wings, warriors with no war. Perhaps the idea of bombing a river was born over a pitcher of beer. A chance to test their talents, avert a flood, put their powers to good use. Back on base, more beer and bravado. No matter the ice hadn't budged.

Richard and Kathy at their Huslia cabin, spring 1976; Kathy pets Shungnak, and Richard holds Seelik. (Nelson archive photo)

"Our ethos," wrote the historian Howard Zinn, "is all that we currently hold to be true. It is what we act upon. It governs our manners, our business and our politics."

Modern society's relationship to nature is carved by confidence. Pipelines, planes, interstates, and bombs are born from the endless accumulation of knowledge and power. Scientists and engineers test and fail, rebuild and learn. There is no problem we cannot solve. No truth we cannot unravel.

The Koyukon relationship to nature is shaped by luck. The gifts of animals and the grace of weather flow through an elaborate culture of humility and restraint. Hunters and trappers test taboos, reinforce and refine the governing rules. There is no success without luck. The world is rich with mystery.

"After they bombed the river," Catherine said, "we had real bad floods for years."

The ice groaned and cracked and ground downriver in late May. Birch buds swelled, staining the hills a vibrant but faint green. Robins and white-crowned sparrows sang near the edges of ponds, covered with lingering rinds of ice.

The last day of the month began with a flurry of snow and ended with intense full sun. That evening, Steven stopped by. "It's time," he said. "Tomorrow we go."

THE FIRST MORNING OF SUMMER camp, the low, fast thrumming of a ruffed grouse lifted Richard and Kathy from sleep. The sound poured through the tent's thin wall, mixing with the songs of sparrows, thrushes, warblers, chickadees, and yellowlegs. With Kathy's head on his chest, Richard lay in his bag and imagined the vibrant curtain of voices shimmering across the vast expanse of boreal forest.

They emerged from the tent to find Catherine blowing life back into the campfire. After a breakfast of hotcakes and roasted fish, they clambered into the Attlas' riverboat to check the net strung through an upstream eddy. They pulled a dozen whitefish and a few pike from the net, then drifted back to camp. Catherine split the fish lengthwise and hung them from poles to dry in the summer sun.

> *In the afternoons we usually worked on projects. Steven cut down birch and spruce trees, which he then split to make lumber for sleds and canoe paddles. He also carved several spoons, using spruce and birch root for the material. Catherine worked tirelessly on birch bark baskets.*
>
> *Evenings were like afternoons, usually devoted to projects or little boat trips. We sat outside and enjoyed the quiet, or the wind, or the birds, the scenery, and the company. The endless light kept us up late, often until 1:00 or 2:00 AM. By this hour the sun was on its way up again, the sky flamed beautifully.*
>
> *Every day was like this, peaceful and pleasant, filled with fascinating experiences, bringing us into closer touch with this country*

and our good friends and teachers. Definitely the best time we've had during our stay in Huslia.

After a week in camp, a passing boat brought news of Chief Henry's failing health. That evening, as the sun dipped through the trees, Catherine poked at the fire and recalled her last visit with the old man. Before leaving town, she'd stopped by to say goodbye. Chief Henry was thin and weak but alert. When Catherine told him she was heading downriver, he got quiet and his eyes drifted toward the window's light.

"I know my time is near," Chief Henry had said, "though I cannot tell exactly when it will come. I have camped many times beneath the spruce trees, roasting grouse over my campfire. There is no reason to pray that I might live on much longer."

In the morning, they broke camp and returned to the village. Richard and Kathy barely recognized their yard. Vibrant clumps of fireweed and grass bristled between the doghouses. A wall of birch leaves circled the cabin. After the long winter of bare, twiggy branches, their yard felt like a jungle, alive with whispering leaves and singing birds.

The vitality of summer contrasted with the quiet vigil centered around Chief Henry's house. A dozen people sat on stumps around a smudge fire meant to keep the mosquitoes at bay. Others stood near a table of food. Most were from Huslia, but some came from neighboring villages to be near the great man in his final days.

Inside, folks sat in silence or whispered to each other around Chief Henry's bed. Bessie's face was tight with pain as she offered sips of water to her companion of sixty-six years.

Richard returned to Chief Henry's house each day. He listened around the fire, added food to the table, and stepped inside to check on the old man.

I have to hold back my tears so often especially when people come to pray for him.

For his part, Chief Henry is dying with enormous dignity. He lies quietly in his bed, occasionally sits up to smoke his pipe or have a sip of water. His mind remains as intact as ever, although he sleeps much of the time. Last night he told Abraham Oldman a story of his youth, his voice just a whisper but his words just as musical as they have always been.

Chief Henry's body is slowly vanishing into itself, but the man remains as big as he ever was. He talks of dying; tells people not to fear his spirit after he is gone, asks that they avoid sadness. "I am ready to leave," he says, "but it seems hard to do."

During Chief Henry's final days, Richard and Kathy packed to leave. They gave away all the dogs except one; Shungnak would stay with them whatever the future held.

Dear Ma and Pa,

True to form, we have boxes for Hawaii, for Fairbanks, for Yakutat, and for Juneau. We're a traveling show, with bits and pieces scattered everywhere ahead and behind us. I can't say how badly we want a home of our own.

In the long light of evening, Richard and Kathy joined the Attlas around a smudge fire in front of their cabin. Catherine pointed toward the wheezy buzz of a redpoll singing from a birch. "That bird is called *k'ilodibeeza*," she said. "It means 'flutters quickly around.'"

A flock of golden plovers zipped overhead. "*Bibidisis*," Steven said, identifying them. "Young women do not say this bird's name because it has *sis* [black bear] in it."

When the sharp three notes of an olive-sided flycatcher called from the river, both Steven and Catherine stopped to listen. "We don't hear that one much around here," Catherine said. "We call that one *duhtseeneeya*. It's hutłaanee to sing that song. I knew a girl who sang that song long time ago and she died of TB the next winter."

The four friends listened to the birds, poked at the fire, and tried not think about how few such moments remained.

In late June, two weeks into the vigil, Richard joined the circle of men outside Chief Henry's house, fluttering birch leaves silhouetted against an amber midnight sky.

After an hour or two, Richard stepped inside to warm up. Chief Henry slept peacefully, frail and small in his bed. Around the room, women whispered among themselves in Koyukon. In a lull between stories, an old woman began to sing. She sat with hands in her lap, eyes downcast. She sang softly to avoid disturbing Chief Henry's rest. The melody changed; other voices joined.

> *The songs were slow, sweet, and exquisitely beautiful. Although I could not understand the words, the plaintive, drawn-out phrases were filled with lamenting and I was sure they were either songs for the dead or perhaps the loneliness of love songs.*
>
> *I wondered if Chief Henry could hear that lovely, sad song. It would be good if he could die with that sound in his ears, something so purely Koyukon, that bore no trace of the modern lifeways twisting the villages today.*
>
> *I stared out the window. The women's enchanting voices filled my mind while I watched the sky fade to deep orange and purple beyond the trees. I thought of the future. I had a clear image of a beautiful green forest, in a time when the Koyukon people were long vanished from this land, and drifting among the silent trees was that beautiful sad song.*

FREE

Richard unclipped the leash.

Shungnak took a few tentative steps on the wide sand beach.

"Go ahead," Richard urged. "You're free."

After a lifetime tethered to either a dog-yard chain or a sled harness, the husky seemed puzzled. She trotted over to a chunk of driftwood, gave it a quick sniff, and looked back. She gazed toward the surf, glancing back once more. Then she took off. She loped across the wave-packed sand, accelerated into a full sprint, and cut a sharp turn. She looped and leaped, sand flying, tail straight back, tongue hanging from grinning teeth.

Richard and Kathy watched, arm in arm. Shungnak raced away to a distant speck, turned, and came tearing back. She wound past their legs and shot off toward the sea. As the dog ran, Richard and Kathy hauled their tent, surfboards, wet suits, and sacks of food to a tangled pile of logs tossed high on the beach by winter storms.

Yakutat is a small village on the Gulf of Alaska coast, a few hundred miles northwest of Juneau. The skies, heavy and dark when they landed at the town's airport early that morning, had brightened to high, thin clouds. After getting a ride to the outskirts, Richard and Kathy hauled their gear down to the beach.They oriented their wall tent toward the surf, upslope from a jumble of logs. They dug a firepit and built a kitchen out of scavenged wave-worn boards.

By the time they kindled the evening fire, the clouds on the far side of Yakutat Bay had parted to reveal the wide fan of the Malaspina Glacier flowing from the slopes of nineteen-thousand-foot Mount Saint Elias.

Shungnak continued to frolic, chasing scents along the tide line and digging holes in the loose sand.

The next morning, after a fire-cooked breakfast, Richard and Kathy wiggled into wet suits, grabbed their boards, and made for a line of waves curling around a sand point. As they waded into the frigid sea, Shungnak paced and whined at the water's edge. They dove beneath the first wave, gasped from the cold, whooped, and dove through the next.

Outside the breakers, they straddled their boards and basked in the wild splendor, admiring the high mountains and the wide empty beach, reveling in this unpeopled stretch of the Pacific Ocean. A large set of waves rolled toward shore. With a few hard strokes, they dropped onto a wave's steepening face and zipped down the rising wall of water.

They surfed until they began to shiver, then rode a final wave to the beach. Shungnak yipped and pranced in greeting. They peeled their suits, stoked the fire to a roaring blaze, and cooked a second breakfast. That afternoon, they explored, moving slow, hand in hand, dinner-plate-sized grizzly tracks weaving alongside saucer-sized tracks of wolves.

Kathy veered down the beach to investigate a half-buried object in the sand and returned with a glass fishing float cradled beneath one arm.

That first day served as a blueprint for their five weeks of surf-camp living. Rainy days, they huddled beneath the tarp, feeding logs into the fire. Dry days, they beachcombed, adding to the growing jumble of trinkets and treasures festooning camp. Every day, rain or shine, amid small waves or muscled curlers, they surfed, immersing their bodies in the lift and fall of the sea.

The intensity of village life faded beneath the steady thrum of waves. No interviews, no late-night journaling, no worries about cross-cultural uncertainties. Richard and Kathy napped in the afternoon sun, ate fire-cooked dinners, stayed curled in the tent until Shungnak whined for them to come out and play. Richard's heart never missed a beat.

Moonlight mushing along the Arctic coast and sunset rides through the boreal forest had formed the richest days of Richard's life. But watching Shungnak's exuberant freedom, Richard knew his dog-mushing days were done. The whisper of runners and the tug of the team formed potent, raw, indelible memories of a life that was no longer his. He didn't want to raise dogs confined to the end of a tether, didn't want to awake each morning in someone else's community.

In Wainwright, Richard's freakish fascination with the natural world had been shared with everyone in town. He yearned for another such community. He ached for Steven and Catherine's connection to place. He envied their days unfolding alongside the graves of their ancestors—every bird, every hill, the ice, the snow, the moose, and the raven all breathing life into Grandfather Olin's stories. Richard craved a home where each pond and stream, every hill and patch of forest, was named and known.

Richard was in his midthirties and still didn't have an address. He owned little more than a wall tent and a surfboard. His parents were now retired and living the snowbird's life, moving between Wisconsin and Florida. His brother lived in Los Angeles. His ancestors' tombstones stood in unknown Scandinavian graveyards.

Richard and Kathy built their last fire on a calm evening in late August. As sparks bristled into the deepening dark, Richard thought of caribou pawing through snow for scraps of lichen, seals scratching through ice for a breath of air, hunters gathered at the coffee shop, sharing news from the hunt. As the fire settled to glowing coals, Richard heard the last song ringing in Chief Henry's ears.

They broke camp the next morning. Gear piled, Richard whistled for Shungnak. It was time to find a home.

HOME

Richard and Kathy hunched against the screech of jet engines as they walked across the tarmac into the Juneau airport. After more than a month on the beach, their senses were soft, vulnerable. The clack and clamor of the terminal chafed and rubbed. In the bustle of baggage claim, they stood against a wall, wishing themselves back to the calm space of wild country.

Their second morning in town, they had breakfast with their good friend Joe Giefer. Joe was on his second cup of coffee when he saw an ad in the Juneau paper. "This one sounds pretty good," he said. "Cabin and boat shed on five acres of beachfront in Tenakee Inlet."

Neither Richard nor Kathy had ever been to Tenakee Springs. Joe described it as a quaint little town. No roads. Just a line of beachside homes centered on a natural hot spring. Fewer than a hundred people in an inlet thirty-three miles long.

"You should call right now," Joe said. "This place will be gone by lunch."

"Call for us," Richard said. "I don't know a damn thing about real estate."

Joe placed the call, posing as Richard. The seller was an older woman now living in Juneau. Richard and Kathy listened to one side of a thirty-minute conversation, answering Joe's written questions with a thumbs-up or a thumbs-down. By the time Joe hung up, there was an earnest money agreement in place. The down payment would swallow their life savings. The rest would be paid over time.

"Get her a check by this afternoon, and it's yours," Joe said. Three hours later, it was done.

Leaving Juneau, the floatplane climbed over the forested expanse of Admiralty Island, crossed the wide stretch of Chatham Strait, and banked into a narrow inlet cutting into the heart of Chichagof Island. The pilot eased toward the water, floats skipping over small waves. The plane slowed and settled. A forest of pilings supported a cluster of dilapidated shacks mixed tight with freshly painted homes. The pilot steered toward a floating dock jutting from the center of town.

Main Street wasn't really a street—more of a wide path between homes. Bicycles leaned on doorways alongside wagons and carts. No space for cars. That wasn't an accident. Tenakee folks liked it that way. Richard and Kathy shouldered their backpacks and, following a hand-drawn map, walked west.

A half mile out, the wide path narrowed, winding through the forest just up from the beach. Smaller footpaths veered off to a few cabins and shacks perched on rock points or nestled into gravel coves. The space between homes grew with their distance from town. A mile out, they found their place.

Bushes pushed against the windows. Weeds overwhelmed the garden beds. A boat shed, built off the cabin's west wall, stood on pilings over the upper beach. Off one corner of the shed an outhouse cantilevered over a steel bucket cleaned out by the highest tides. The cabin itself was a one-room affair—the kitchen sharing space with a bed and couch. Gravity-fed water piped from a nearby stream dribbled into an old enamel sink. Woodstove, musty bed, doghouse for Shungnak, wringer-washer tub, kerosene lamps, smokehouse, root cellar, and quiet.

They explored all afternoon, picking through tools and treasures cluttering the cabin and shed. In the evening, they walked the beach. Salmon jumped from a mirror sea, the slap of silver bodies loud in the still inlet. The whistling yodel of an eagle rang from a tall spruce. Richard and Kathy turned in circles, trying to take it all in.

Their beach.

Their home.

Before sleep, in the glow of a kerosene lamp, Richard wrote his parents.

Dear Ma and Pa,

The place might not appeal to everybody, but we are crazy about it. Our little cabin is right on the water, big windows open to the view. It's situated on a narrow shelf of flat ground between the beach and the forested slope behind.

The house is quaint to say the least. It is very weathered, but not rotten anywhere. The interior is really cute, and not "weathered" at all. It is jammed full of stuff so we find ourselves needing virtually nothing to move here. It would be fantastic if you could come visit us here next summer.

Kathy worked long hours reclaiming the garden in anticipation of spring planting. Richard sharpened an old crosscut saw he found rusting in the boat shed. He cut spruce poles, built a sawhorse, and began filling the woodshed. They cured a batch of salmon in the smokehouse, picked blueberries for jam, baked muffins in the wood-fired oven. They kayaked around the inlet, meandered up rivers, marveled at bears and whales.

Their neighbor Ole Thorgaard, a retired carpenter, often walked the forest path between their homes. Ole helped prioritize repairs: fresh tar paper and cedar shakes, new metal roofing, double-pane windows to replace drafty single-panes, insulation and paneling.

They closed the door on their dream home and left Tenakee in mid-October for an anthropology conference in Fairbanks, resource management meetings in Juneau, and then a seaside rental in Koloa, Hawaii, where Richard and Kathy spent the winter summarizing their Huslia research for the National Park Service. They returned to Tenakee in April.

Dear Ma and Pa,

Got here yesterday. The place is just as peaceful and beautiful as ever. Things have just started getting green—little hint of leaves, sprouting grass, neighbor's flowers starting to bloom.

There are a million jobs to be done. Most pressing is the garden—it's planting time about now. We'll have to get the soil turned and get stuff in the ground.

The barge with our building materials is due in any day so we'll be able to get to business with the construction projects. Got to get new pilings under the place. Then there's windows, insulation, and roof. It'll all be hard work, but lots of fun too.

In early May, the explosive exhale of the season's first whale filtered into Richard's dreams. He shook Kathy awake, and they stood outside, half-naked, and watched the humpback cruise past. Days later, sprigs of lettuce sprouted in the garden, and the first melodic notes of a hermit thrush joined the morning chorus of ruby-crowned kinglets and song sparrows.

"My strongest memory of that homestead," Richard recalled, "is silence. No generator. No chain saw. No water pump. Just hand tools. Our bodies and minds were immersed in that place. We'd never dreamed of finding such a perfect home."

But in early June 1977, the rhythmic chop of a helicopter cut the inlet. A second helicopter replaced the first. All that day and the next, a string of choppers carried slingloads of supplies to the inlet's far shore. The high whine of distant chain saws filled the space between flights.

"They're gearing up to log Crab Bay," Ole told them. "They'll harvest Kadashan Bay next."

Dear Ma and Pa,

Just learned the sad news they are going to log the hillside straight across from us. Broke my heart. I guess they're going to get all the other bays eventually too. My first inclination is to sell and move somewhere else, but it doesn't take long to realize that wouldn't solve a thing. We live in a time when there is no escape from the ravages, and so we just adjust to the destruction around us. This is the wrong time in history to care deeply about nature.

The weatherman had called for a stretch of sun. Richard gathered his tools and ripped into the cabin roof. Too old for ladders, Ole shouted advice from the yard as Richard threw down sheets of rusted metal and lengths of old rafters. Destruction of the old roof took a day. Construction of the new spanned a week. With Ole's guidance, Richard framed a dormer over a second-story loft.

The sun held. Richard screwed fresh metal onto a dry roof. Kathy climbed the ladder to help lift the window into place. It was late evening, loggers done for the day. Richard and Kathy sat on the roof in the gathering dusk, basking in the pride of accomplishment. That night, they slept in the new loft. Their window framed a view across the inlet where tree tips blended like feathers, texturing the far hillside with a soft pattern of unbroken green.

The next morning, Richard swept sawdust from the rough plank floor while Kathy washed the breakfast dishes, in anticipation of a visit from the Attlas. When Richard had invited them to visit, he didn't think they'd actually come. He knew Steven and Catherine rarely left the river basin of their birth. He knew the airfare—four flights landing in six communities—would burn a big chunk of their annual income. The return letter saying they'd be there in June was a sweet surprise. Richard hung the broom on its nail behind the door, carried an armload of tools to the boat shed, and readied the skiff to meet the midday plane.

After a life among the stunted, sparse trees of the boreal forest, Steven and Catherine were astonished by the rain forest. They marveled at the soaring trunks, thick moss, wide armored leaves of devil's club, and abundance of bell-shaped blueberry blossoms. They studied deer tracks and squirrel trails, bear scat and otter dens. Listening to the chorus of birdsong, Catherine smiled. "Listen to that robin. They sing so different down here."

The Attlas had no stories, no experiences, to prepare them for the rushing breath and arching back of a humpback. And then seals, porpoises, and the high black fins of orcas—the seaside show had no end.

A riverman to his core, Steven watched the drift of kelp and birds, the swirl of water over stones. When the tide turned and the water slowed, stopped, and flowed the other way, it turned his world upside down. It was like the sun stopping midday and descending back toward the east. During his week in the inlet, Steven never missed a chance to watch the saltwater river change directions.

Toward the end of their stay, Catherine found Richard cutting firewood behind the cabin. She smiled, listening to the rhythmic zip of the crosscut. As he worked, metal teeth singing with each push, each pull, she sang along, her voice pitch-perfect and in sync with the saw.

After a few rounds, Richard paused to catch his breath.

"Back home, it's more and more motors all the time," Catherine said. "You're living the way we used to."

Fresh log on the sawhorse, Richard cut again.

"It's good to see you here," Catherine said. "And it's so good to hear that sound."

Richard later described Catherine as an "absolutely brilliant human being. A genius by any standard. She and Steven embodied a depth of

Richard, Kathy, Catherine, Steven, and Shungnak in Tenakee (Nelson archive photo)

knowledge on par with the greatest intellects from any culture." But that afternoon, in the quiet shade by the woodshed, Catherine was far more than a genius. She was a friend, a mentor, a mother. Her simple words about the saw's song came as a subtle, profound blessing.

Too soon, it was time for them to go.

They said their goodbyes on the floatplane dock, the pilot standing by.

"Wait. Almost forgot," Richard said. He grabbed a bag of fish fillets pulled from the smoker that morning, and handed them to Steven. "Food for the way home."

A light rain dimpled a calm sea as Richard and Kathy boated back to the cabin. Beads of moisture glistened on tips of Shungnak's fur and wisps of Kathy's hair. Richard tethered the skiff and caught up with Kathy alongside the garden. They nibbled a few greens, talked about the dry dirt needing a good rain.

Kathy pulled weeds. Richard grabbed a shovel and started digging a hole for a fresh piling under the boat shed. The chores helped fill the space emptied by the Attlas' departure. The low clouds and descending mist muffled the distant chain saws.

DYNAMITE

Three evenly spaced whistles rang across the inlet—a loaded stretch of quiet followed by a concussive explosion. The blast rolled across the water and echoed off the mountains. Dynamiting began shortly after the Attlas left. "Rock quarry," Ole said. "They gotta build a lot of road to get the trees out. Whistles let a guy know it's time to get to safe ground."

Seven days a week, the whine of saws mixed with the grumble of trucks and the detonation of dynamite. In late June, Richard looked out the new loft window and noticed a small break in the distant canopy of trees. Just a subtle tear, a few feathers missing from a duck's breast. Forty acres out of a thousand, but the scar demanded attention. A dozen times a day, his eye wandered to the distant blemish. It was like an infection swelling across his own body—something he didn't want to see but couldn't help but study.

> *Dear Ma and Pa,*
>
> *I'm not adapting well to the constant noise—the helicopters, the heavy equipment, dynamite and saws. I love this land so much. I can hardly tolerate to see the destruction of its beauty and tranquility. We just weren't meant to live in this age, where people act as if all the earth is just made to gratify men and machines.*
>
> *We'll either have to adjust or move further out into the boondocks.*

Later that month, Richard paddled across the inlet for a closer look. He stashed his kayak in the rye grass and pushed his way through the thick fringe of alder bushes. He stood for a moment in the dark forest. It was evening, the workday done. No saws. No trucks. Uphill, a squirrel chattered.

In the beachside forest, everything looked fine: a familiar mat of moss spreading between soaring trunks of spruce and hemlocks, deer trails winding through a tangle of blueberry and mensiezia. A strip of sky shining through the trees marked the road's edge a quarter mile uphill. Richard stepped from the forest onto the gravel. Tall trees hugged the curving roadbed like canyon walls.

When he heard the whine of a rig, he jumped into the forest and ducked behind a tree. A dusty pickup rumbled past. Richard wasn't breaking any laws. He was in the Tongass—a national forest owned by all Americans. He had just as much right to be there as the men in the truck. Still, he felt like an intruder, a spy, as if logging had transformed the forest into a foreign country.

Richard stepped back onto the road and hiked around an uphill curve, toward a wide swatch of sky. In the months listening to the trucks and saws, he had imagined a scene reminiscent of the selective logging he'd witnessed in the Midwest—a few stumps scattered through an intact forest. Nothing he'd seen or read or imagined prepared him for his first encounter with a clear-cut.

To his left and right, a wall of intact forest towered over an apocalyptic world. The moss, the bushes, the trees, every scrap of living tissue, lay shredded and torn. A lifeless jumble of stumps, branches, and shattered trunks stretched upslope. Broken bodies in a quiet field.

Years later, Richard said that coming upon a clear-cut always induced anxiety: "It's the same foreboding I sometimes feel in depths of sleep, when a blissful dream slowly degenerates into a nightmare; I am carried helplessly along, dimly hoping it's only a dream, but unable to awaken myself and escape."

Richard paddled home in the deepening dark.

Before this, the noise of industry had been mostly an irritant to him, an acoustic intrusion. After his paddle, the whining saws and rumbling trucks cut beyond the loss of his personal dream. The wholesale slaughter of trees echoed the devastation of bison and tall-grass prairie, the slash of the pipeline and riverbank carcasses of headless moose.

Dear Ma and Pa,

The more I think about the future, the more I become convinced that we're right to work toward a self-sufficient homestead. Hard to tell where fuel and food are going to come from in the years ahead—best to prepare for the worst. I hate to think what today's newborns will experience.

Don't get the idea we aren't loving our life here. On the contrary, we love it with incredible intensity. Can't get enough of it.

Kathy's long hours in the garden paid off, with more radishes, lettuce, onions, and beet greens than they could ever eat. Bushes within a stone's throw of the house produced buckets of salmonberries and blueberries. Each day, more salmon pressed into the inlet and nosed up streams.

On a late July evening, Richard and Kathy stood on their beach and gazed across a placid sea. Everywhere they looked, silver bodies popped up. Freeze a moment of time, and there'd be hundreds of fish suspended in midair. Thousands along the inlet's full length.

But every night, before sleep, Richard sat on the edge of his bed and studied the mountain's growing scar. He pictured the day's work—the fallen trees stripped of limbs, bucked to length, yarded, stacked, and trucked to the beachside sort yard. He imagined the panicked squirrels, the frightened birds, the scattered deer. He fantasized about a midnight raid: sugar in gas tanks, engines drained of oil, anything to stop the destruction.

He also knew he didn't have it in him. Richard knew he'd find common ground, over a burger and a milkshake, with any of the truckers or fellers working across the inlet. Meeting violence with violence was not his way. What then to do with the grief and anger? How to pick up the pieces of a shattered dream?

PART III

Island Years

HAVEN

Smooth dark waves steepened over a rock reef. The crests crumbled, spitting white water and spray that glistened in crisp fall sun. Clad in a thick neoprene suit, Richard rode an offshore current back out to the lineup, surf crashing just off his shoulder. Beyond the reef, he straddled his board and sat up to catch his breath—not a boat or body in sight.

Foam on the back side of the surf flashed in bright contrast to the island's jumbled black shore. A wall of coniferous trees lifted from the beach, cloaking gentle slopes in unbroken green. The symmetrical cone of a snow-clad volcano pushed up through the forest to dominate the sky line. The year's first snow dusted the volcano's rim.

Richard shook his head and grinned at the day's beauty. He threw his head back and let loose a joyous whoop swallowed by the wide Alaskan sea. Feeling the muscular lift of a big set, he lay on his board, kicked hard, and dropped into the mysterious world of wind and water. He cut to the right, skittering across the moving wall of liquid glass.

In the trough below, water boiled over bedrock. Foam crashed at his heels. All thought—the bite of betrayal, the fear of loss, the ache for home—vanished beneath the swirl of sensation.

Exhausted after a few hours, Richard stroked for shore. He steered past a ragged headland toward the arc of a sand cove. Shungnak pranced on the beach and barked. Richard pulled off his flippers and walked through the surge. He sat on a drift log as the dog licked salty water from his face. Richard peeled off his thick mitts and nestled chilled fingers into her warm fur.

Five years had passed since Kathy left. For almost every one of those almost two thousand days, he'd kept Shungnak in arm's reach. She was there when he'd first learned of Kathy's lingering love for another man. The faithful dog stayed close through that first winter alone. She lay beneath the table through the long solitary hours.

Shungnak was in Seattle as Richard sought the closeness of friends. She traveled back to Tenakee with him, sniffing along the beaches and trotting the trails while Richard decided he'd keep the place, try and live there on his own. And she was there as he journaled by candlelight.

> *There are dangers out in the wild places, perhaps more than in a town; but I see no emotional dangers in the wilds. I have struggled with the fear of the emotional peril human relationships entail. Nature is my haven. Alone in the wild country, I find release from the fear—hidden beneath the shelter of trees, sustained by the stability of rocks, moved by the certain power of the sea. The only clear and definable malevolence I can see resides in the human realm.*

Shungnak suffered the heat and confinement of California when Richard took a short-term teaching job at UC Santa Cruz. And she was there when Richard met Nita.

Nita worked in administration on the Santa Cruz campus. Richard was renting a room from her office mate Val. With Val's encouragement, Nita just happened to stop by the apartment on a Saturday morning. Richard was just in from surfing, hair still wet. "He was genuine," Nita recalled. "He had integrity, humility, and such an obvious respect for the people and places of Alaska."

Nita was gentle and inquisitive. She glowed with a soft-spoken kindness. In the following weeks, they shared dinners and walks and sleepovers. But for Richard, the loss of Kathy was still close, still raw. Fall from a cliff one time, and you tend to stay away from the edge. Besides, Richard would return to Alaska; Nita was rooted in California. He had little interest in fatherhood; she had a four-year-old son.

When Richard finished teaching in June, just a month after their first date, he said goodbye and headed north. A few days later, Nita's phone rang. The call itself was a surprise—she thought she'd never hear from him

Richard with a large halibut caught in Tenakee Inlet (Nelson archive photo)

again. Even more surprising was the invitation to bring Ethan to Alaska for a visit.

Tenakee worked its magic on Nita—catching fish, tending the garden, baking cookies in a wood-fired oven, watching her boy toss rocks into the still inlet. Her first visit widened into a vision of a new life.

For Nita, a permanent move from California was hard. She was a single mom, grateful for all the time her parents spent with her son. But the decision to go to Alaska with Richard was easy. She trusted him. She knew she and Ethan would be all right.

Life in Tenakee was serene, simple but, ultimately, unsustainable. Neither Richard nor Nita could earn money there. Ethan didn't have playmates. And the ragged edge of the widening clear-cut chewed a growing hole in Richard's heart.

They imagined the ideal community: bigger than a village, smaller than a city; a town with job options and a wider circle of friends for Nita; decent schools and a tribe of kids for Ethan; nearby wild country and surf for

Richard. They decided on Sitka, population eight thousand, nestled on the outer coast of Baranof Island, a day's boat ride from Tenakee.

Richard shut the door and carried the final armload of gear to the beach. He tightened the straps of Ethan's life vest and lifted him into their skiff. Shungnak leaped aboard, dripping seawater across Nita's lap. Richard stood ankle-deep in the inlet, one hand on the boat. He turned toward the cabin, gazing at the garden beds and the dormer window.

"Come on," Ethan said. "Let's go."

Gripping the gunwale with both hands, Richard pushed off and jumped in.

WATCH AND WONDER

The narrow, paved streets of downtown Sitka wrap around either side of an old Russian Orthodox church. As Richard wandered past on his first day in town, a chortling croak made him stop. On either side of a white cross rising from the church's roof sat two black ravens. The birds murmured, clicked, cackled, and purred. Richard swung his binoculars from beneath his arm and stepped from the curb for a better view. The ravens flared their neck feathers, fanned their tails, bowed up and down. They hopped through the points of the cross, from peak to tip and back, babbling and crooning.

When the birds flew off, Richard lowered his gaze. A half dozen people had gathered, curious to glimpse what riveted his attention. He sheepishly tucked his binoculars under his arm and hurried on.

A block later, he was drawn to a window display of books. Inside, a jovial elf of a man tended the counter. Don Muller, owner of Old Harbor Books, was a fan of Richard's writing and was delighted to welcome him to town. They stood at the counter, chatting for the better part of the afternoon. Before leaving, Richard asked if Don knew of any conservation groups working in town.

"SCS—Sitka Conservation Society," Don replied. "Our next meeting is tomorrow night." He scrawled an address on a scrap of paper. "We'd love to have you."

The meeting was in the home of Lee Schmidt, a local physician. Don was there along with half a dozen others. Seated around Lee's kitchen table, Richard snacked on cheese and smoked fish while he learned about local logging; in a 1959 contract signed with Alaska Pulp Corporation, the federal government had promised the mill a fifty-year supply of trees. In exchange, Alaska Pulp committed to build and operate two mills, one in Sitka and another in Wrangell, about 120 miles southeast.

Plates of food got replaced by stacks of maps, existing clear-cuts in red, proposed cuts in black. "The primary goal of SCS," Don explained, "is to slow the flow of logs from the Tongass. Five hundred million board feet of old-growth every year! For what? Disposable diapers and toothbrush handles? The Tongass timber program operates at a net loss of ten million dollars per year. This is taxpayer-funded destruction of a public forest."

The group was on to coffee and ice cream when Richard shared his experience of paddling across Tenakee Inlet and his first encounter with a clear-cut. "I don't know if we'll ever learn to treat this world with respect," he said, "but thanks to you all, there might be a few trees left standing while we try." He took a bite of ice cream. "Sometimes it feels like the only thing we've learned since exterminating the buffalo is how to kill things faster."

Don set down his cup. "I move that Richard Nelson be appointed to the board of Sitka Conservation Society." The motion drew a hearty round of ayes, and the meeting adjourned. Richard slipped into his coat with no idea his tenure on the SCS board would span decades.

Richard and Nita rented a small apartment not far from the Russian Orthodox church. From a desk in a tight living room corner, Richard worked on his fourth book, *Make Prayers to the Raven: A Koyukon View of the Northern Forest*. His previous ethnographies, *Hunters of the Northern Ice* and *Hunters of the Northern Forest*, had focused on the techniques, skills, and specialized knowledge unique to Inuit and Gwich'in people. *Make Prayers to the Raven* moved beyond lifeways to explore the Koyukon's spiritual relationship with the world:

> *As I was living among the Koyukon people, nothing struck me more forcefully than the fact that they experience a different reality in the natural world. . . . For the Koyukon, there is a different existence in the forest, something fully actualized within their physical and emotional senses, yet entirely beyond those of outsiders. . . .*
>
> *Although I had lived for several years among other native Alaskan peoples, I had never found access to this element of their lives. . . . The great difference in our points of view was something I had been*

prepared for intellectually; but I was entirely unprepared for it emotionally, unready for the impact of living it.

Fundamental assumptions I had learned about the nature of nature were thrown into doubt. I must emphasize that I underwent no great conversion and emerged no less an agnostic than before. But now I had to face an elemental question: . . . Is there not a single reality in the natural world, an absolute and universal reality? Apparently the answer to this question is no.

After a long day of writing, Richard joined the other SCS board members around Lee's kitchen table. Discussion centered on the next proposed clear-cuts just north of town. That night, Nita breathing deeply beside him, he couldn't sleep. He thought about the morality of his own culture—a culture of never enough. Never enough gold, never enough land, never enough oil, never enough trees. He thought of Chief Henry on his deathbed, expressing gratitude for all the nights he'd camped beneath spruce trees, roasting grouse over the fire. Food and warmth—Chief Henry never asked the land for more. His desires meshed with his needs. He had no assets, no need for a will. Richard realized that Chief Henry had left behind something of far greater value:

We often remember ancient or traditional cultures for the monuments they left behind—the megaliths of Stonehenge, the temples of Bangkok, the pyramids of Teotihuacan, the great ruins of Machu Picchu. People like the Koyukon have created no such monuments, but they have left something that may be unique—greater and more significant as a human achievement. This legacy is the land itself, enduring and essentially unchanged despite having supported human life for countless centuries.

The intact landscape caused Richard to reflect on his own culture's relationship to wild country.

The fact that Westerners identify this remote country as wilderness reflects their inability to conceive of occupying and utilizing an environment without fundamentally altering its natural state.

In early fall, mountains lost in low clouds, a letter arrived from David Brent, the editor handling *Make Prayers to the Raven* for the University of Chicago Press. As an academic press, the publisher sent all manuscripts out for peer review, and Brent had sent Richard the critiques. A few of the anonymous reviewers were adamant that personal reflection had no place within ethnography. "The author's journal entries," declared one reviewer, "do not belong in the manuscript."

Richard called David seeking advice. "Sure, you could cut all contemplative stuff," David said. "Cut it all and the reviewers would have the book they want, but it wouldn't be the book anthropology needs."

Buoyed by Brent's support, Richard sat down to write the book's epilogue. He thought back to a crisp winter evening, mushing his team through a stand of tall spruce outside of Huslia. He recalled a raven flying overhead, disappearing and reappearing behind the treetops.

> *Certainty is for those who have learned and believed only one truth. Where I come from, the raven is just a bird—an interesting and beautiful one perhaps, even an intelligent one—but it is a bird, and that is all. But where I am now, the raven is many other things first, its form and existence as a bird almost the least significant of its qualities. It is a person and a power, God in a clown's suit, incarnation of a once-omnipotent spirit. The raven sees, hears, understands, reveals . . . determines.*
>
> *What is the raven? Bird-watchers and biologists know. Koyukon elders and their children who listen know. But those like me, who have heard and accepted them both, are left to watch and wonder.*

BECOMING WORTHY

Binoculars pressed tight to the jet's window, Richard traced the volcano's smooth, forested slopes down to the jumbled, tortured shoreline. He picked out a beach where he'd recently camped with Nita and recognized the nearby muskeg where they'd hiked. He scanned over thousands of acres of yet-to-explore forest, miles of unwalked beaches.

A line of swells swept in from the horizon, wrapped around Kruzof Island's southern tip, and outlined the black basalt in a wreath of white foam. He spotted a few reefs where he'd already surfed and noted several others with good potential. The plane rose suddenly into clouds. The island lingered in his mind long after it vanished from view.

AFTER THEIR FIRST SUMMER, RICHARD and Nita knew Sitka was home. They bought an old house a few miles south of town, on the shore of Jamestown Bay. Ethan's room, not much bigger than his bed, was just off the kitchen. Richard and Nita slept beneath the low, sloping attic ceiling. An old boathouse, perched on pilings over the rocky beach, served as Richard's office. A window framed the forested shores of nearby islands and the distant rise of Mount Edgecumbe.

Richard's first job at his new desk was lead author and editor of an anthology titled *The Athabaskans: People of the Boreal Forest*. Through the wet days of fall, as Richard thought about dry interior forests, a new project shimmered in the rain.

On New Year's Day 1984, Richard opened a fresh spiral-bound notebook. On the top of the first page he wrote:

RAIN: I

A SITKA JOURNAL

JAN. 1, 1984 Overcast, rain (sometimes heavy) high ca 50, low last nite ca 40, strong SE wind

The New Year rushed in powerfully on a storm, our first in weeks. The house was warm this morning, although the stove had been out all night as usual. I looked out the window and noticed the snow had vanished, leaving puddles amid the brown grass. A robin picked at the ground, apparently oblivious to the rain.

When I let Shungnak out I was almost startled by the warmth, after weeks of temperatures between +10 and +35. No wonder—the thermometer read 50. I called the news upstairs to Nita and she was as delighted as I was, but the winter-lover Ethan was disappointed. He thrives on snow and seems oblivious to the chill. Nita and I would rather have clouds and warmth than sunshine and cold.

The casual tone of the opening paragraphs belies the role the Rain Journal would come to play in Richard's life. For each of the next 5,110 days Richard, without fail, made time for this journal. Through bouts of flu or times of stress, days at home or weeks on the road, alone on Kruzof Island or in town with friends, he filled page after page, notebook after notebook, with reflections, observations, confessions, struggles, and insights. In the privacy of his handwritten entries, he experimented with lyricism and coaxed his growing dreams into the light of awareness.

After long days editing the Athabaskan anthology, Richard turned to the journal:

The job itself is rather mundane. A straight ethnographic treatment allows little room for creativity or imagination. I long to write a book away from notes and references, drawing on memory, feelings, and invention. Buried under the stilted language of science, all but the faintest flicker of humanity is extinguished. The people described in such traditional ethnographies would function as machines, without the lubricants of emotion, individuality, and estheticism that are vital to the life of every culture. How anthropology, a discipline devoted to understanding the human species, became so removed from whole, vast dimensions of human experience is beyond understanding.

Although confident Sitka was the right choice, that first winter in their new home was hard on everyone: Ethan worked out fresh friendships in a new school, Nita sought meaningful employment, and Richard struggled with insomnia that had begun when Kathy left.

Lately I've had much trouble sleeping; just another bad spell in seven years of chronic insomnia. I work all day in the half-fog. My body is heavy and weak, and sometimes I feel sick. I spend the day dreading the time I'll have to go to bed. Anyone who has never suffered insomnia could never understand how awful it is.

I wonder again, why do I love this cursed cold and dreary place so much? Some people have cruel lovers, but they find beauty enough to tolerate the abuse. I must be one of those. . . . Were it not for the wildness I would never live outside the tropics; but I cannot live without wildness. . . . If you must have wild country you must have something to drive everyone else away. You hang on, teeth gritted and fists clenched, biting into the frigid wind and cringing against the chill rain.

Richard held tight, editing the anthology through the gray days of winter. On a calm March morning, the sun evaporated all possibility of desk-bound chores. Richard loaded wet suit and Shungnak into the skiff and skimmed west over a wide, smooth swell.

There were no known surf breaks, no other surfers in Sitka Sound. Richard was a pioneer, nosing between reefs, easing through kelp beds, exploring the collision of ocean and rock.

Splay-legged in the pitching boat, Shungnak whined in protest as Richard mused about the waves thundering over offshore reefs.

I have never looked at waves without trying to understand why they break as they do—what tricks of water and rock and wind have shaped them—yet I still understand next to nothing. I could watch a million waves leaping and flinging like fire, and the last would be no less thrilling than the first.

Finding a promising break, Richard put Shungnak ashore, anchored the skiff, wiggled into his wet suit, and slipped into the sea. Stroking for the lineup, he watched steep dark walls rise against the reef's far edge. The waves gleamed like jewels: liquid, moving, alive. He quickly dropped in.

I have that wonderful sense of euphoria and contentment. It is one of the most satisfying sensations I've ever known, one that affects my body and mind together. The best analogy is sexual, though it is misleading because there is nothing really sexual about it. Still, the sense of physical and emotional fulfillment is parallel, the peace and calm, the warmth and happiness and love.

I have never surfed without thinking aloud "I love you, Ocean." I have never surfed without saying thanks to the great mass of water that brings such ecstasy and wonder.

The continued warmth of spring added weight to the anthology's looming deadlines. In April, as bees worked blossoms just outside the office window, Richard forced his attention back to the project.

This is a magnificent day—sunny, warm, gentle wind, and the water glittering. It takes a supreme act of self-control for a free man to stay inside but, somehow, I overcome the cravings of every molecule in my body. I hunch over the papers at my desk and press on like a slave in chains. Let the birds and bushes soak up the brilliance and warmth; I am a human; I have papers.

But how can I complain? No one keeps my hours and I am as free as anyone can hope to be in this orderly society, where few people can schedule their own lives, where most file into their cages in the morning and file out again at dusk. How can the Native Alaskans look at the way we live and not feel pity? Yet each generation moves increasingly toward joining our morning processions rather than trying to help us out of this mess. How strange we send them our teachers, when we should beg them to send their teachers to us.

One evening in May, while Richard washed the dinner dishes, Nita turned on the television in their small living room.

"Come in here, Nels," she said. "You've got to see this." Richard walked over, drying his hands. He immediately recognized the rhythmic snap of skin drums and Iñupiaq voices. And, drawing close, he recognized the faces of dancers moving across the screen.

> *As I watched the dance and listened to the songs my mind slipped away to Wainwright, now twenty years since I first arrived there. I heard the same songs in the crowded hall, saw these same men singing with the drums, felt the floor shake beneath pounding feet. My mind wandered off onto the great jumbled sweep of sea ice, drifted at dusk to the edges of leads where I hunted with these same men, now singing in my television.*
>
> *I look from face to face, and see how age has changed you all. And how you changed my life: I can scarcely fathom all the ways. I catch myself sinking to sadness for the times now gone, the times we spent together, and think of the gifts you gave me. You made a man of me, put me through your school of life and tested me to the last fiber of my body and soul. But most of all you changed my way of looking at the world, of thinking about nature and loving it.*
>
> *You taught me that a man can hunt and love the animal he stalks, the one that gives himself so that he can live. Take an animal with humility and show it gratitude, leave aside the false dream of domination, speak to the thing you have killed and to the ones you leave unharmed. Nature listens.*
>
> *The dance quickens and the drums pull me back. I leave the past behind, though it will live in me forever. There are ways of becoming worthy of the lessons taught you, most of all by passing them along.*

THE ISLAND CALLS

After a restless night in a stuffy hotel room, Richard woke early to the sounds of horns and sirens. It was mid-May. Richard was in Anchorage for a week of meetings. As he hurried past a city park, the three-note song of a white-crowned sparrow cut through the traffic's din. The sound transported Richard to spring camp on the banks of the Koyukuk River.

He recalled Catherine Attla singing back to a bird—*yee sik its'ee tee tłot*—and then telling the story of how, in the ancient world, a man had starved on his way to spring camp. When he died, his shell necklace transformed into the sparrow's white crown. He flew up into a spruce, and the birds have his song and spirit still.

Richard pressed on, now late for his meeting. That night, back at the hotel, he pulled out the Rain Journal.

> *This afternoon I watched a raven fly above distant houses and thought of the few animals who choose to remain where the forests have been—sacred emissaries in a profane world. They sing out like holy men chanting on dusty corners, where papers fly in cold winds and businessmen hurry past with briefcases, paying no attention.*
>
> *My mind wanders back beyond a century, to a time when the land beneath these houses, lawns, and streets sustained a different life, was forest and muskeg, and hunting ground. Moose snap through the willows; lynx prowl the shadowed rims of bluffs; a great gray owl planes down on silent wings to snatch a vole from the forest*

floor. I see an aged spruce and wonder if hunters passed beneath it, carrying sinew-backed bows and copper knives.

We have buried the holy places beneath laundromats and service stations, pawn shops and playgrounds. We have shut sacredness away inside churches and temples, where the images are touched by no wind or sunlight, and where no animal save mice and rats will ever move. But for me, holiness is not inside the churches, but in the pigeons that nest in the parapets and the crows that fly in the clear above the steeples. I hear nothing of the sacred from the man who preaches from an altar, but the raven's call rings out with it.

All week, Richard sat in crowded rooms with windows that did not open. The task at hand: reduce the complexity of subsistence living into simple graphs and figures, for legislators too busy to fully understand lifeways different from their own. The meetings were exhausting yet clarifying.

By the end of the week, I am feeling altogether removed from the processes of science and politics that now govern our relationship to the land. I realize that over recent years, my work has taken on more and more distance from authentic experience. But this is all changing now; I am turning back toward the source.

At week's end I withdraw from a research project I had planned to do. I am filled with freedom and elation.

Back in Sitka, Richard and Nita indulged in a late, lazy morning. Eventually, Ethan rousted them from bed, restless to get out in the warm sun. Compared to Anchorage's late spring, Sitka was bursting with eager green leaves and strident singing birds. After a simple waffle breakfast, Richard sat to catch up with a backlog of desk work. A hummingbird buzzed past the window. Freshly opened salmonberry flowers beckoned like pink jewels beyond the glass. Shungnak nosed his elbow and whined. Richard set down the unopened mail and reached for his running shoes.

Loping along the familiar seaside trail, he and his dog moved beneath the descending notes of a hermit thrush and into the rising swirl of a Swainson's thrush. Rounding the point near the river mouth, the chattering melody of a ruby-crowned kinglet mixed with the chaotic chorus of crows.

Shafts of afternoon sun flashed through towering spruce and patterned the forest trail with bars of light. Richard's feet felt light on the spongy path, his laboring lungs and pounding heart a welcome respite after days of travel.

Puffing at the end of the run, Richard glimpsed a brown creeper scratching up the trunk of a dead spruce. The bird pecked at the crumbling bark with its slender, curved bill, then darted after a small gray moth.

Watching the twitchy spark of life, Richard realized he'd never heard a creeper's voice. As his own breath settled, he wished for the bird to sing, wished to deepen his acquaintance with this tiny neighbor.

My education is so incomplete; I feel so ignorant. Watching the creeper I realize what I want. I want to educate myself about the woods and the shore with the natural world as both subject and teacher. I want to leave aside the filters of other men's minds—the middle men who carry their own ideas and preconceptions into classrooms far removed from the subject.

What insights are only attainable through the senses? I'd like to run the experiment: Spend a year in the forest university, studying with as much discipline as I would on a campus bound by walls. I want to twist direct experience together with insights gained from indigenous teachers.

Right now, anthropology is behind me, perhaps forever, and the thought fills me with excitement. At last I'm coming home. The truth of it is so clear I wonder why it took so long. Never mind; soon my daily footsteps will be in the deep moss where they belong.

The island calls.

BLANKET OF WORDS

Richard eased himself into the crowd packed between the shelves of Old Harbor Books. Gary Snyder was a big name for a small town; like so many of his neighbors, Richard was eager to hear what the famous poet had to say. He had first encountered Gary's poetry on the UC Santa Barbara campus in the late sixties. Gary's recent writings about wildness resonated with Richard.

It was a humbling surprise when, weeks earlier, Gary's letter had arrived in the mail. The poet wrote of his own background in anthropology and expressed admiration for Richard's work. In closing, he said, "I'll be in Sitka soon. I hope we have the chance to meet."

The morning after the reading, the two writers boated over a smooth swell toward Kruzof Island. They anchored on the south shore, shouldered day packs, and pushed into the woods. They skirted the edge of a muskeg, the air still, the snow-clad slopes of Mount Edgecumbe stark against a high gray sky. A junco family ticked and fluttered near the base of a stunted pine. The faint perfume of bog orchids mingled with the pungent odor of Labrador tea. Shungnak sniffed along the edge of a shallow muck-bottomed pond. A water strider ran to the far side, needlelike legs dimpling the glassy surface.

Richard and Gary moved at an easy pace, admiring plants, trading ideas about living close to the land, exploring practical problems of surviving as writers. They lunched on a bedrock rise beside a twisted bonsai pine draped with strands of delicate green lichen. The bog stretched in all directions, quiet but for the rhythmic huff of a raven's wing, the chattering of a red squirrel, the whispered roar of distant surf. Gary commented on the terrain's uncanny resemblance to the Africa savanna—spreading shore pines reminiscent of acacia trees, the looming volcanic slopes an echo of Kilimanjaro.

After lunch, they followed deer trails winding through the dim light of heavy forest. Shungnak snuffled near the base of a hemlock tree. Richard dropped to his knees to see what she'd found and lifted a deer bone from deep in the moss. Gary also knelt, and the two men combed fingers through the feathered green, discovering ribs and vertebrae, a scapula and a femur. Gary then lifted a disfigured lump of bone—a foreleg, broken and then healed. Richard brushed off bits of dirt and hemlock needles.

The deer must have lived a long time after its terrible injury. Long enough so the fragments knitted themselves together, as if liquid bone had seeped into the wound and solidified as a porous, bulging, convoluted mass. Though gnarled and misshapen, the fused bone seemed almost as strong as a healthy one. But the deer's leg was considerably shortened and had a hollow ivory splinter piercing out from it. As we turned the bone in our hands, I marveled at the determination of living things, and of life itself, to carry on, to mend, and to become whole again after being torn apart.

Afternoon slid into evening as the two men sat by the tree and talked, the deformed bone on the ground between them.

I have never met anyone whose ideas so keenly reflect my own; but Gary has thought them out in far greater detail and merged them into a real philosophy. I have only edged myself that way and begun making it a life pattern. His ideas about humanity living with nature shine out to me like a clear sunrise. He has brought together and made order of so many questions that now guide my own life—what is the spiritual orientation that fosters long-term survival of both humans and their environments? How do we heal our broken relationship to the natural world?

TWO MONTHS AFTER GARY LEFT, Richard was on the island alone. He packed a sheet of metal roofing up the beach, holding it over his head like an oversized hat. At the forest's edge, he dropped his load, crawled through a low gap in the prickly wall of spruce branches, and pulled the roofing in behind him. He snaked the metal between tightly spaced trunks toward a structure tucked deep in the trees. The writing shack was tiny, eight feet by ten feet, just big enough for a bunk, a woodstove, and a single wooden chair. The roofing and stove came from town. The walls, floor, and bed were cobbled together from boards and sticks tossed ashore by the waves.

Richard chose the spot with an eye for obscurity and isolation. Interlocking branches made the cabin invisible from both planes and boats. Local knowledge and intense grit were required to successfully navigate the rock-riddled entrance to the nearby anchorage. The storms pounding the exposed stretch of the south Kruzof shore meant Richard would have the nearby surf breaks and hunting muskegs to himself.

Richard screwed on the last piece of roofing, smeared some tar around the chimney flashing, and kindled the first fire in his home away from home. From the beach, he watched the thick smoke waft and eddy out of the trees on a stiff September breeze. The shack was a visible manifestation of a slow-cooked dream: within the snug shelter, his private journal would grow into a public book, a poetic narrative integrating lessons from other cultures with insights gleaned from the island itself.

The next day, Richard returned home and put the finishing touches on an application for a three-year fellowship with the Kellogg Foundation. He was painfully aware his hunger for intimacy with the island would not help with house payments and grocery bills. He'd already applied to the Guggenheim Foundation and the Alaska State Council on the Arts. He was nervous to call this third attempt done, reluctant to sign his name and seal the envelope. He fretted over each word, desperate to inspire the review committee to fund his plan.

> *If I were to propose a course of study on an accredited campus or seek an apprenticeship with a renowned professor, the funds would flow. Somehow, I have little faith that "scientists" will see the point of spending a full year with an island as my teacher—observation and intimacy my only tools, the Earth herself my only reference.*

RICHARD SQUIRMED IN HIS SEAT. A bead of sweat formed behind one ear. Carolyn Forché, the evening's first reader, turned to a new poem. The audience sat rapt, silent, attentive. It was the third evening of Sitka's annual weeklong writers' symposium. Richard was the next presenter. The room was packed.

He'd fretted all day about sharing passages from the Rain Journal. Over the course of the winter, he'd spent weeks alone in the writing shack, scratching out the most intimate, lyrical passages he'd ever penned—words shared with no one, not even Nita. He scanned the room. Everyone appeared transfixed by the agony bleeding from Carolyn's poems. Richard contemplated running home and grabbing something, anything to replace his thin pages.

Long, loud applause. Carolyn was finished. No time to run.

John Straley, novelist, poet, and one of Richard's closest friends, sidled over in the brief break. A veteran of public readings, John knew what Richard was up against.

"You're screwed, man. She killed it."

"Thanks, John."

"I'm just kidding. You got this. Go get 'em."

Richard settled at the podium, looked at the expectant faces. He began with the easy stuff—paragraphs from his most recent books, *Shadow of the Hunter* and *Make Prayers to the Raven*—published passages about someone else's life. And then it was time.

All the candlelit nights in the shack, all the hours immersing his senses in sea and forest, his aching desire to inspire humility and love for the natural world—it all felt like he'd been singing in the shower at the top of his lungs and suddenly the curtain was lifted and he was onstage, naked and off-key.

A sudden twinge of fear as I turn to the journal. I feel like everything rides on this, the whole course I've followed these past two years and the course ahead as far as I can see. Barry Lopez and Robert Hass are in the room. Other writers. My friends. They're all listening.

Richard introduced the evening's final passage by sharing his long-held dream of touching a wild deer. He recounted naïve fawns prancing just beyond reach and how he'd begun to think his dream was far-fetched, perhaps even foolish. He then described a November hunt and his encounter with a doe bedded in a snowy, sun-soaked muskeg. Cold rifle clutched at his side, he eased close, pausing long minutes between steps. He huddled behind a slender pine, just a few yards from the doe. A large buck sauntered into the open sunshine. Richard read on:

> *When he moves up behind her she stands quickly, bends her body into a strange sideways arc, and stares back at him. The buck moves to the warm ground of her bed and lowers his nose to the place where her female scent is strongest.*
>
> *Inching like a reptile on a cold rock, I have stepped out from the tree and let my whole menacing profile become visible. I am a hunter hovering near his prey and a watcher craving inhuman love, torn between the deepest impulses.*
>
> *Drawn to the honey of her scent, the buck steps quickly toward her. The doe turns away from him and walks straight toward me. There is no hesitation, only a wild deer coming along the trail of hardened snow, the trail in which I stand at this moment.*
>
> *My existence is reduced to a pair of eyes; a rush of unbearable heat flushes through my cheeks; and a sense of absolute certainty fuses in my mind.*
>
> *I am struck by how gently her hooves touch the trail, how little sound she makes as she steps, how thick the fur is on her flank and shoulder, how unfathomable her eyes look. I am consumed with a sense of her perfect elegance in the brilliant sun. And then I am lost again in the whirling intensity of experience.*
>
> *She never pauses or looks away. Her feet punch down mechanically into the snow, coming closer and closer, until they are less than a yard from my own. Then she stops, stretches her neck calmly toward me, and lifts her nose.*
>
> *There is not the slightest question in my mind, as if this was sure to happen and I have known all along exactly what to do. I slowly raise my hand and reach out.*

And my fingers touch the soft, dry, gently needling fur on top of the deer's head, and press down to the living warmth of flesh underneath.

Richard glanced up from the page; the audience was tight and attentive. He read on, still nervous but buoyed with the new certainty that the writing was good. He finished the passage, the doe exploding across the muskeg in exquisite bounds and Richard basking in sincere applause. As he stepped from the podium, he was wrapped in a blanket of kind words from just the people he'd dreamed to hear them from.

RAIN JOURNAL

June 25, 1985
Partly cloudy, W breeze, ca 60/ca 50

Some surge, apparently a moderate SW swell outside

This is page 1000 of the journal, a year and a half since it started. I've never read through it, but know I've gone through a great change in this time. This journal is the focus of a whole reorientation in my work life—in my whole existence, really—bringing the things I do into harmony with the person I am. Although I have great insecurity about my ability to make a career of this, of writing about our relationship to the world, I have ultimate determination to give myself entirely to the effort. And in the process, I have reached an inner comfort I haven't known since I was too young to worry about such things.

Aside from devotion to family and friends, this writing and the experiences behind it are the center of my life now. The new work has brought an intense excitement to daily existence for me, and given me that unexcelled energy that comes from trying to do something you aren't sure you're capable of. But, most of all, it has taken me closer than I've ever been to the nature I've always loved. And, because of that, the work can never fail no matter what the writing ever means to anyone else.

June 27
Partly cloudy then clear, SE wind then W pm, ca 58/ca 45

Thrush songs from the backyard forest and thicket are a constant sweet perfume these days. Bumblebees are thick in the flower beds;

they give a voice to the flowers. The humpback whale decides to try Jamestown Bay this morning and apparently finds a rich repast. We watch for an hour as it bubbles and bursts up on the calm water right off our beach. I wonder if any of our neighbors are watching television while this great show of life goes on in the full power of multidimensional reality right outside their windows.

What a great tragedy that we so often trade vicarious experiences for the chance to pursue real ones. The fullness of our senses is wasted on the artificial imitation of the world, most of their dimensions lost in the flatness of postcards and television screens. We are trapped in a dim half-life while the living world bursts forth outside our windows, in our own backyard and in the wild fields beyond.

June 29
Shoals Point to St Lazaria
Clear, NE breeze then SW pm, ca 70/ca 50
No swell

I crave more time on the island, but suppose it will have to be done alone. No one else has the patience for long, quiet watching. I feel like I could stay here for days, just watching, waiting. Let the island show itself to me in its own time. Ordinarily I am hyperactive and impatient, but wild nature so captivates my whole attention that I forget the passage of time. I can do it for hours—just watching things, feeling very occupied and busy. None of my friends are affected this way, so I'm always aware of them, hurried, my attention half consumed by worrying about them. They are bored, anxious to be somewhere else. Much as I thrive on their company, especially having Nita along, I realize the real work in all this has to done alone. The only true, absorbed watching is a solitary thing.

July 7
Overcast, haze, W wind, high upper 50's, low ca 52
Small surge

Much as I love freedom in the outdoors, and much as my perspective darkens under the stress of work, I still have a deep need to feel I'm "accomplishing something" every day. What a burden that is—mentally and physically—when most of my activity is outdoors.

The key is combining the two imperatives. Be in nature and accomplish something professionally through it. Living in the villages up north has been a way to do that, and no doubt this explains why I did it so much. But there are other stresses there, ones I can no longer tolerate. So now I pursue this new plan, to live intensely outdoors here and write about the experiences. Perhaps this will solve my Protestant dilemma.

July 10

Overcast, light rain, S/SW wind, ca 60/low 50's

Slight surge, small surf visible on Low Island

Reading accounts of early visitors to the great prairie I marvel that it ever existed at all. Where they saw the land buried between its horizons with buffalo migrating through undulating oceans of grass as high as their shoulders, today we see a checker board of cornfields haunted by starlings. I am grateful that some who saw it left their accounts so we can know what we have lost.

I am grateful too, because knowing of that loss I can love this place all the more. By good luck or accident, humankind has not yet exterminated the natural inhabitants of this shore. I feel like an explorer of the last century here, at the leading edge of the onrushing cataclysm, recording my experiences of a living world that may soon be destroyed. The plow has already touched earth here, this time a lumberman's saw laying waste to great forests . . . and not even the decency of a crop planted in bare soil to cover it.

Sometimes I step outside myself and look at these words from a time not long in the future. I see myself a figure in a faded daguerreotype, standing amid a wild landscape long gone from the face of the earth. And I wonder if such a world as this could ever have existed. It makes me live this miracle all the more intensely, love it all the more, and wish that bittersweet hope that the future might never be.

I look up at this very moment, full of sad longing, and through the notch between two islands I see a whale's back arch on the channel waters, its flukes rise against the dark mountainside and sink into the pure silence below.

I give thanks for the world, for the gift of dying before it is gone, and for my eyes that live now and can see.

July 11

Sitka–Shoals Point–Redoubt–Sitka

Thickening OC, drizzle and haze PM, mostly calm, low 50's/ upper 50's

Some surge, S swell outside, surf 5-6' at The Wall

It's one of those days when paddling out is almost as much fun as riding, the waves are so beautiful to watch. I play word games inside my head to describe them: gleaming green glissades, spinning sparkling spirals, the singing ringing water. I lay my face in the water to feel it, to touch it in some intimate way, to be as thoroughly in it as I can. I am overflowing with pleasure and gratitude. Is this what a hawk feels like, playing where the gale sweeps up the mountain?

July 16

Beaver Point–Sitka

Overcast on the outside/clear elsewhere. NW wind (strong pm), ca 65/ca 50

S swell 3–4 ft. along south Kruzof, especially west of S. Lazaria

Nita and I are up early after a poor night's sleep. Too many nocturnal beetles in the tent, and some mosquitoes. I get up at 3:30, distracted by a sound like someone hitting a dry log with a stick, to stare into the dawn for the source. It is seals slapping the water in their dawn frolics.

Boating home we fall in line with two humpbacks—a mother and her calf. The two whales blow together, close enough so I can watch them angle down beneath the water heading our way. Nita shouts out. I look down and see it too. The mother whale stretches through the green shimmering depths twenty feet beneath us. I can see the details of her broad, spatulate snout, tubercled and blotched, with the over-slung jaw clamped up against it. Her flippers are against her body, not extended. Her calf glides just beneath her.

That moment is frozen in my mind, fixed like the image of the sun when you close your eyes after looking toward it. The two great bodies submarining so close we could dive down and touch them.

And I know I understand absolutely nothing. Never have I been more forcefully aware of the vast distance between myself and a whale than during these moments of close proximity. We could have touched, we could have looked into one another's eyes, but we know nothing of the separate worlds we live in and the separate worlds our senses bring to us.

March 25, 1986
Snowline halfway down the mountains.

High overcast with sunbreaks, SE breeze, few showers, ca 40/ ca 58

Slight surge, no whitewater

I'm getting deeper and deeper into the process of writing. It's taking me over. The work is so exciting that my heart sometimes pounds while I'm thinking and writing. The subjects and events are so much a part of my daily life that they seem ordinary and I wonder if anyone will be interested. People like Barry Lopez and Gary Snyder help me recognize what the things I'm talking about mean in a larger perspective.

I've asked Nita to read any letter from the Guggenheim Foundation, and if it's negative just put it away until this writing is done. I don't want any big emotional setbacks right now, with my confidence pretty shaky and my courage at a premium. I still believe the absence of formal support or endorsement for this work makes me all the hungrier and determined. If it stays this way I'll have only the Island and my friends and teachers to thank, which might be ideal.

July 25
Sitka–Redoubt–St Lazaria–Sitka

Overcast, rain and drizzle am, high OC pm, W breeze, low 50's/ ca 60

SW swell to 4' on the sound

We climb up near the west end of St Lazaria and make our way out to the cliff top above the murre colony. Sitting in the grass to escape the stiff westerly wind, we watch gulls soaring on the updrafts, at eye level and so close we could talk to them in a normal voice. I'm captivated by the exquisite beauty of these gulls. I'm especially taken by the way the feathers on top of their wings and shoulders ruffle up in the vacuum that lets them stay aloft. Easy, drifting flight, like flecks of bark afloat in rushing water. You can't help envying them, wanting to abandon this complex life and clumsy body, launch out from the precipice and spend your days hanging from feathered sails in the ocean wind.

July 30
Shoals Cove–Sitka

Partly cloudy AM then overcast by noon, rain PM tonite, strong SW wind, SE PM, ca 62/50

Very small swell; slight surge at Shoals Cove, surf ca 1' at Low Island

Closeness is the sacred power I seek. My amulet comes by moving within the touch of eyes, mingling scents, reaching out with my fingers toward feathers ruffled by the same wind gust that surrounds us both. I ask no blood from the eagle, no feathers. Only look at me; let me hear breath wheezing through your beak, the ticking of your quills, the scuffling of your feet on dry bark, let me feel the rushing air beneath the rhythm of your wing beats.

August 21
Overcast, rain, SE wind then SW pm, ca 50/ca 60

Slight surge

Each lap around the trail offered glimpses out to the split horizon of sea and high arc of Edgecumbe, feathered against the clouds. The sight never fails to make my heart swell, the pure pleasure of being close to such perfection and beauty. If anything remains from me when I die, it should be left on Kruzof Island, where my bones can crumble into that soil and my atoms mingle with the muskeg and moss and lava. There could be no better afterlife for me than to become part of the island—to run through the blood of deer and flow in the muskeg ponds, to fly over the high peaks in the feathers of eagles, to soak with rains into the deep sphagnum. Raven, take my eyes and see with them. Wind, take my breath and sing with it. Hemlock, take my flesh and grow tall with it. Island, take my soul and twist it together with your own.

PAST GIFTS AND FUTURE POSSIBILITIES

Robert Rose stopped by on a cloudless, calm morning. The peak of Edgecumbe cut a dark edge against the pale-blue sky.

"Well, what do you think?" Robert asked.

"This is it. Let's do it," Richard replied.

For years, the two friends had surfed and hunted in the mountain's shadow. Around many a campfire, they'd mused of someday standing on the volcano's rim. The unexpected stretch of September sun was the year's last chance.

By midday, they buzzed across placid water, jacketless in 70-degree warmth. Boat anchored, they stuffed food and gear into backpacks and slipped into shorts for the long hike. Shungnak pranced and whined, eager to get going.

It was Robert who suggested they spend the day in silence. A potter, boatbuilder, and practicing Buddhist, he viewed the hike as a walking meditation. For Richard, the forced silence seemed odd, but after a few hours the quiet felt like hunting—stalking an experience instead of deer.

Breaking into the bright warmth of the first muskeg, the trail became a furrow of mud, every step a squish down and a hard pull up. The trail gradually steepened as they wound through ravines and emerged onto an open ridge. Soon the mountain loomed as a great wall, each step more up than forward. Pausing to catch his breath, Richard clung to the mountain and soaked in the view.

In all the land surrounding this mountain, there is only this one narrow thread of trail. The fragment of the island I will eventually come to see contains infinitely more dimensions than I could ever comprehend, even if I could be here every day of my life. Even on this island, which is so small it disappears from most maps, I am only a fleck on the surface and a flicker in time. I am no more significant than the bubble that rises on a muskeg pond, shines for a moment, then pops without leaving the slightest ripple. I mean no more here than the buzz of a hummingbird's wings or the click of a pine needle falling onto the tangled crowberries.

The thought comforts me, releases me from the dreams of self-importance, relieves me of the need to be anything except here and alive and watching the great world that contains me.

From the volcano's rim, Richard watched the sun drop into the wide stretch of ocean. Turning east, he saw the shadow of Mount Edgecumbe spread across the waters of Sitka Sound, then slowly rise up the pink-tinged mountain slopes above Sitka.

Richard pulled on a sweater in the deepening chill and studied the lights of town through binoculars. He could just make out the tiny twinkle of his own home. He felt a pang of loneliness, wishing Nita and Ethan could share this magic evening on the mountain. He gazed out over the Pacific's deep glow and stared into the Kruzof crater's unblinking black eye. In the meek light of a headlamp, cold fingers barely able to grip his pencil, he wrote:

If I could list my first ten choices of places on earth to be right now, they would all be somewhere on the rim of this volcano.

Richard unrolled his sleeping bag in a small swale. Tired from the hike, he quickly drifted off, but each time he awoke and saw the crater's dark rim slicing through a sea of stars, his heart pounded with excitement.

The sky is thick with stars, like a handful of flour pitched up and set aglow. I watch the great arc of the Milky Way, hour after hour, like the hands of a celestial clock as the earth revolves beneath it. Sleep seems irrelevant, almost wrong, when I think of how precious my time is here.

He awoke to Shungnak's wet nose on his cheek. He sat up—the crater rich with color, the eastern sky pale blue and amber. Far below, muskegs outlined with dark trees rose up the mountain's flanks like an exquisitely quilted skirt. The Pacific, drawn with silent lines of ocean swell, kissed the far reach of western sky. That moment, awakening on the island's peak would, in the next eighteen months of desk-bound writing, become a beacon, a burning memory lighting the darker moments of the work.

Richard spotted a tiny figure moving along the far rim and noticed Robert's empty sleeping bag. Walking briskly in the morning's chill, Richard scurried across the crater's rim and met Robert heading back to camp. Silence over, the friends babbled about dragonflies, horned larks, sunsets, and starlight.

Home from the volcano, Richard's first inclination was to grab a sack of groceries and return to the island. Instead, he took a clean sheet of paper and penned himself a sign. With a thumbtack in each corner, he pinned it above his desk window:

IF PAST GIFTS AND FUTURE POSSIBILITIES MATTER . . .
STAY HERE AND WRITE

He then turned to his journal.

It's over now. I've left. The "research" is done. I've gone from the Island to my journal to my mind.

Every day I have a strengthening sense that I have to do this well, that so much depends on it, that I could never get any deeper into my soul than this work lets me go, that I owe so much to its source and want so much to say something, that I can scarcely keep this intensity inside any longer, that I am aching to get these words written at last.

In writing ethnographies, Richard, scissors in hand, had cut his field journals, paragraph by paragraph, and pasted the fragments onto five-by-seven notecards. He'd organized cards by subject: BEAVER, TRAPPING compiled alongside BEAVER, NATURAL HISTORY. MOOSE, STORIES filed beside MOOSE, HUNTING. The hands-on cutting and pasting took months, but the resulting shoeboxes of cards defined the structure and content of the final books.

When Richard declared the island "research" done, he had nearly two thousand journal pages. Instead of using scissors and glue, this time he created a written index. He planned to give it a few weeks. It took the better part of a year. The final index contained 159 alphabetized subjects—starting with ABUNDANCE and ending with WRITING. They filled twenty-two tightly typed pages. A small sample:

BEAR

encountering tracks,
encounters with,
sharing the Island with,
danger of,
at sperm whale carcass,
spring behavior of,
active in,
returning skull of,

DEATH

forest graveyard,
and life beyond death,
of things in nature, (eagle), (cliff-fallen deer), (animal's mistakes), (falling to predators)
of sperm whale,
of hunter's lost deer,
and autumn,
Nita's mom, (last conversation), (funeral time), (Christmas after)
and sadness,
SEE ALSO: Nature, Predator, Salmon

DRAGONFLY

occurrence of,
description and behavior of,

NATURE

connectedness to, (city dwellers), (observing), (Gaia), (Gaia, the eagle's eyes), (where all comes from), (wild vs. store foods), (eyes), (earth shaped to whatever lives), (Gaia/boundaries)
and artificiality of boundaries,
loving and knowing,
closeness to/alienation from,
knowledge of,
cruelty/reality of,
asking nothing of,
relationship to,
human place in,
and human creations,
sacredness of,
prayers to,
as place of security, (source of sanity), (trusting nature)
being renewed in,
austerity in Alaska,
proximity in Sitka,
in the city, (crow in truck), (waxwing on wire)
controlling, (gardens)
destruction of, (midwest landscape), (loss of prairies), (loss of animals), (humanized Wisconsin lake)
recovering the earth,
protecting,
SEE ALSO: Earth, Place

SHUNGNAK

guiding me in darkness,
her first freedom,
importance to me,
gets hurt,
SEE ALSO: Hunting

Midway through indexing, in October 1986, the latest and last letter of regret arrived in the mail—this one from the Alaska State Council on the

Arts. Richard filed it alongside rejection letters from the Kellogg Foundation and the Guggenheim Foundation. The clean sweep nourished his ever-present self-doubt but also carried relief.

> *I have this odd sense I've escaped the last risk of external involvement in the Island experience. I've come this far on my own, it seems best to keep the whole thing between me and the Island. Never should have taken the risk in the first place.*

The STAY HERE AND WRITE sign was not an absolute moratorium. Richard occasionally slipped to the island to fill his senses and hunt. At his desk, the words came sometimes in a trickle, some days a welcome gush. In the woods, it was a luckless fall, deer elusive and wary, hour after hour, hunt after hunt. The Koyukon notion of luck seemed to follow him through the forest. Had his writing offended the deer? Was it a mistake to pin the intimacy of the hunt onto the indelible, lifeless page?

It was odd to think seeing animals was beyond his control, that all he could do was be there, step quietly, make himself available should the deer choose to appear. Standing in the shadows staring across a frosted and still muskeg, he recalled being on the edge of a lead on the frozen Chukchi Sea with his old friend Tagruk.

> *I remember my friend whispering near the edge of the black, steaming water. "Seal head . . . seal head . . . come, seal . . . " Twenty-five years later, I understand in a much different way the emotions behind calling an animal to yourself.*

Back in his office, Shungnak at his feet, Mount Edgecumbe framed in the window, his days filled with the peace of purpose.

> *I feel myself becoming steadily more intense about the work, more anxious to get at the creative stuff of writing, and more confident about the rightness of the path I'm on.*
>
> *I pick up the binoculars several times each day to look at the Island, as if I'm reaching over to touch a person I love and to be*

touched in return. I look at the Island and think, "I'm doing this for you."

This is such a wonder, loving a place as if it is another person. How strange to have reached my age without ever living in a way that allowed it. I have certainly read of it, I have lived with people who premised their whole existence on it. But I was not raised in a culture or community that allowed it to happen to me.

There is only one way to know this love: Go to the land, become rooted in it, and let the love grow through intimacy and completeness of experience. It's no different than loving a person. A real love emerges through intimate connectedness, not by reading about a person, not by looking at someone's photograph in a book, not by the occasional passing glance or fragment of conversation. Real love emerges over time spent in intimate togetherness, founded on mutual dependence and reciprocity. As it is with the community of people, it is with the larger community of nature and place.

FLOOD TIDE

"Thoreau wrote about Walden mostly after leaving it," noted the American essayist Robert Finch. "The best accounts are always separate finished episodes—Melville at sea, Whitman in the first flush of self-discovery, Hemingway at war—rather than the impossibly disparate grist of one's current daily existence."

After three years of journaling and months of meticulous indexing, Richard took Finch's words to heart. In January 1987, he bought a one-way ticket to Seaside, Oregon, packed his computer and surf gear, and bid a tearful farewell to Nita and Ethan. He rented a quiet apartment a block from the beach and poured ten to twelve hours a day into the book. First, he mapped the narrative arc. Second, he assigned indexed scenes to specific chapters.

Finally, he was ready to write. Across the top of a blank computer screen, he tapped:

Chapter One

The Face in a Raindrop

With a deep breath, he traveled north to Lavine Williams's low-slung cabin on the banks of the frozen Koyukuk River. "A good hunter . . . ," the old man said as they drank steaming black tea, "that's somebody the animals *come* to."

The words rolled smooth, building toward a rain-drenched island scene that embodied the essential question beneath all the years of yearning and learning.

> *The nettling rain seems drawn against me. It drips from my face, wrinkles my hands, seeps down inside my boots, soaks my hair, runs down my neck, penetrates my heavy, sodden clothes. . . .*
>
> *I could grumble about the rain and the discomfort, but after all, rain affirms what this country is. Today I stand face to face with the maker of it all, the source of its beauty and abundance, and I love the rain as desert people love the sun. I remember that the human body is ninety-eight percent water, and so, more than anything else, rain is the source of my own existence. I imagine myself transformed back to the rain from which I came. My hair is a wispy, wind-torn cloud. My eyes are rainwater ponds, glistening with tears. My mind is sometimes a clear pool, sometimes an impenetrable bank of fog. My heart is a thunderstorm, shot through with lightning and noise, pumping the flood of rainwater that surges in my veins. My breath is the misty wind, whispering and soft one moment, laughing and raucous another. I am a man made of rain.*
>
> *At this moment, there must be more raindrops falling on the surface of the island than there are humans on earth, perhaps more than all the humans who ever lived. I've thought of raindrops as tiny and insignificant things, but against the scale of earth itself, they're scarcely smaller than I am. On what basis, then, can I consider myself more important?*

Each day, Richard broke from his desk for a surf check. Winter storms lined up an endless rhythm of waves, some mushy and unformed, others too steep and intimidating for surfing. But most days, he suited up and paddled into the sea, ripping along the edge of his abilities.

Like the waves, the writing varied. When the words flowed, Richard's self-confidence grew. Other days were awash in doubt. By early February, he'd crafted his way through the first chapter and into the second, where he told of a stormbound night on the island.

I listen to the steady throb of surf that resonates through the trees, and the chatter of raindrops on the tent wall. My heart is torn between the island and home. Born into a culture that keeps the worlds of humanity and nature apart, I am always close to one love but longing for another.

Richard missed Nita's warm smile and quiet reassurances. Six weeks into his writing retreat, he surfed a final wave, packed his gear, and returned to Sitka.

After a sweet reunion with Nita and Ethan and a welcome run with Shungnak, he set up the computer on his familiar desk in the old garage. Through the growing days of spring, the long stretch of summer sun, and the bluster of fall storms, Richard's mood swung with the ebb and flow of words, productive euphoria followed by derailed depression. By year's end, he'd finished chapter seven.

January 1 dawned clear and cold, the ocean's swell too small for surfing. Friends dropped by; an afternoon ping-pong tournament stretched into a communal dinner. Before sleep, Richard sketched his dreams for the coming year.

(1) Continued happiness with Nita and Ethan, most of all; a contented home, let Shungnak stay healthy. (2) Finish "The Island Within," hopefully by May 1; and have it accepted by a publisher. (3) Freedom! After May 1 or whenever the book is done, I want to pour myself back to the outdoors—work around home, lots of time on Kruzof, surfing as often as there are waves, and new explorations. (4) Solve the imperative to find some income, hopefully from writing. A book advance? A grant or fellowship? Magazine articles? Eskimo ethnography? Anything but a job . . . (5) Find escape from the constant little injuries that have limited me—get rid of the tendonitis so I can run, play soccer, play volleyball, play more water polo and surf, surf, surf. (6) Time with Mom and Dad, hopefully a nice visit with them here. (7) And finally, all I really want is a simple life, here in Sitka, peaceful and private, made rich with love and daily adventure, closely touching wildness.

Richard wrote eight to ten hours a day. He forced himself to take weekends and holidays off. The self-imposed deadline came and went. The whales and birds returned. Late May, the final two chapters surged like a flood tide.

The air was thick with mist when Richard slipped from bed on the morning of June 9. He'd slept poorly, excited to feel the end of writing drawing near. The day was a mind bender of blending scenes, cutting words. That evening, he printed a copy and sat down for the final read-through. When he reached the last line—*Ethan, joyous and alive, boy made of deer*—he could only read aloud "Ethan . . ."

The words caught in his throat. He cried. What would readers think? Should anything so private ever be made public? He knew only this: He'd held nothing back. He'd loved a place as fully as he could and felt the land's love in return.

The next morning, Richard meandered through the yard. He touched the almost-ripe salmonberries and admired the goatsbeard coming into full blossom. He breathed the pungent perfume of rhododendrons and delighted in watching heavy-bodied bees ambling through stalks of purple lupine. It felt like his first close look at spring: everything more vivid and saturated than he remembered.

Nita and Ethan and their dear friends John and Jan Straley gathered for hot dogs and games. Eagles swooped down and scooped talonloads of herring. Seals swam just offshore. The clink of horseshoes mixed with the cries of gulls.

As the sun sank toward the island's summit, Richard walked away from the fire and laughter and gathered his thoughts.

> *Every time I look at the Island I get this wonderful feeling inside. It's partly gratitude, partly excitement about getting back there again, and partly a sense that "We did it," as if the Island and I have finished a collaboration. After these 18 months of writing, I have a much stronger bond with the Island than ever before. Like any love, this one is evolving from hot, flaming, romantic intensity to a deeper and more enduring relationship. I hope it goes on and on.*

SHUNGNAK'S GRAVE

Out of the tent before dawn, Richard perched on a drift log beneath an August sky ablaze with stars. Shungnak settled at his feet. Surf crashed and rumbled in the dark. As the sky brightened, Richard slung his rifle and headed for the woods. The old dog rose on stiff legs and followed, struggling to keep pace as they clambered up a steep bank.

At the edge of the first muskeg, Richard leaned his rifle against a slender pine. Wispy pools of fog twisted through low swales and drifted over still black ponds. Spiderwebs glistened with beads of dew. A thick-bodied dragonfly, translucent wings jeweled with moisture, clung to a spike of crowberry. Beyond it all, Mount Edgecumbe rose like a wall, high ridges blushing with dawn's first light.

After the long siege of writing, the island's beauty leaped through his senses like fire. Gratitude pooled, ran down his cheeks. Richard made no effort to wipe away the tears. By the time he moved again, the fog was gone, the muskeg vibrant with deep shadows and sharp sun. The dragonfly had floated away, wings rattling briefly against blades of grass.

He called Shungnak with a gentle "Psst." She wandered across the muskeg, ignoring a chorus of his increasingly loud calls. A twinge of irritation melted with the realization his companion was growing deaf. So much change between one hunting season and the next. How many more island mornings would he share with the old sled dog?

The next day, back at home, Richard and Nita quietly probed a far different reality. The day the finished manuscript was mailed, they were both unemployed. They had $1,740 in their shared account. Nita's job as a researcher in a law office had ended. To keep his focus on the island experience, Richard had turned down a dozen jobs.

A month later, Richard Jordan, chair of the anthropology department at the University of Alaska in Fairbanks, asked him to teach a course on the lifeways of Athabaskan people. Despite dwindling funds, Richard couldn't bring himself to commit to what he called a "standard subject, standard approach, exams, papers, the full catastrophe." He drafted a counterproposal, outlining two seminar-based courses: Anthropological Writing, and Nature and Mind. Jordan agreed.

With classes scheduled to begin in the New Year, the family fretted over how best to split their lives between Sitka and Fairbanks. Nita, interested in enrolling at the university, wanted to go. Ethan, a preteen embedded in a tribe of friends, wanted to stay. John and Jan Straley solved the dilemma by offering to house Ethan while Richard and Nita went north.

Splitting the family made Christmas together all the more precious. After pancakes and presents, the three of them took a long, leisurely stroll up a steep mountain road, light snow swelling to thick flakes as they climbed. Near the top, Nita pulled a hacky sack from her pocket.

> *For the next hour or so, the three of us play in the whirling snow. Such a rare pleasure, this frivolity on the mountain top, surrounded by whirling flakes and sagging boughs. Shungnak watches from a distance. She looks so beautiful to me, here in the snow, as if we're magically transported back in time, to our mushing days. I feel such tenderness every time I look over to where she waits. I'm so grateful to be alive, in this place, with this little family I so dearly love.*

A RAVEN PRATTLED AND CROAKED from a birch on the edge of the Fairbanks campus. Richard walked closer, pushed back his hood, squinting against sun reflected off fresh snow. The bird spread his throat feathers and babbled on. This interior raven, voice magnified by the superchilled air, spoke a language similar yet distinct from the birds chortling in the seaside spruce trees of Richard's home.

Fairbanks days flew by, packed with prepping lectures and grading papers. In addition to teaching, his book's birth commanded attention. Following protracted weeks of agonizing silence, Jack Shoemaker, editor with North Point Press in Berkeley, California, called in March with an enthusiastic offer to publish *The Island Within*.

When Richard hung up, he was surprised by his calm, cool response to what should have been rapturous news.

> *I wish I felt more. I guess the real excitement comes each day when the book is being written. I remember the creative energy raging inside me, words and sentences flying through my brain, each evening's thoughts on the next day's work. It's the writing that thrills me. Everything else is a more subdued sense of reward or satisfaction, a high that follows a daily creative rush.*

Susan Bergholz, a New York–based literary agent, negotiated the contract and coordinated publicity efforts with the publisher. Richard was grateful for her business sense and her confidence in the book's success. Later that month, Susan called with big news: *Life* magazine wanted to run a full excerpt, photos included. This time, Richard's reaction wasn't so subdued.

> *Never, never did I expect something like this; Audubon or Outside Magazine, yes, but not LIFE. The Island book has been driven by a dream of reaching a larger audience, not just for personal gratification and success, but also because I believe so deeply in what the book is about, believe in the importance of nurturing a different way of seeing nature. This isn't a writing exercise for me, it is a cause I believe in like nothing else. I see the world I love so utterly being laid to waste, and all I have to offer is a voice.*

To preserve the island's obscurity, Richard had fictionalized the geography throughout the manuscript. Every stream, summit, cove, and inlet got renamed on a map burning in his mind alone. He insisted the book's jacket cover make no mention of Alaska and refer to his home as "a small town on the Pacific Northwest coast." So when Susan read a draft of the *Life* article and saw Tongass National Forest referenced in the opening sentence, she knew there'd be trouble. Richard was devastated.

My heart sinks. I have no choice but to refuse, and to back up that refusal with a willingness to withdraw the article. I undertook this thing with a personal vow to respect the anonymity of place, and I must stick by it. Whatever I've been given has come to me within the bounds of that promise. If I broke the promise now—for the sake of promoting myself or my work—I would violate these gifts, violate myself.

Withdrawing the article turned out not to be necessary. The magazine agreed to excerpt the book without referencing the Tongass. (Richard later said he would not fictionalize the geography if he were to publish *The Island Within* today. He came to believe that nourishing the love of specific places is the best way to ensure their protection.)

Mid-April, Richard and Nita awoke to the welcome sweep of a warm wind. Juncos scratched for seeds at the base of a birch, the click and tick of their voices cheery after winter's prolonged quiet.

In the late afternoon, they took Shungnak for a walk in brilliant sun. A half mile from the house, she slowed, panting heavily. She collapsed in the soft snow and lapsed toward unconsciousness. Richard and Nita dropped to their knees on either side of the ailing dog. After a few minutes, Shungnak opened her eyes and licked Nita's hand. Richard ran for the car, and they soon had her back at their apartment, where she drank, ate a bite of meat, and curled into a ball.

Spring rushed on. Mushy snow gave way to squishy mud. Robins and white-crowned sparrows sang from high branches. Lines of geese and flights of cranes filled the evening skies. Oblivious to it all, Shungnak slept, endless hours inert on the living room rug.

Classes ended in late April. Nita packed, eager to rejoin Ethan in Sitka. Richard had to stay and finish grading papers. The night after Nita left, Shungnak wandered the house, struggling to stand, whining, and bumping against walls and furniture. No amount of petting or soothing words could ease her agitation. Through two more sleepless nights and restless days, Richard tried to calm his companion before coming to a sobering conclusion.

I silently beg for death to come on its own. Strange, I've seen so many animals die at my own hand, but this one is almost impossible to face. The time may have arrived when I owe her death, and when I must accept my own pain in exchange for ending hers.

Richard carried Shungnak into the clinic the next morning and placed her on the steel table. The vet shaved a bit of hair and inserted a needle into a nearly collapsed vein, Shungnak tight with fear, Richard gripped with grief. The blue drug slid from the clear syringe, and the dog's head rested heavy and still in Richard's arms. Richard slumped, burying his face in Shungnak's fur. For a long time, he couldn't move, couldn't stop the sobs, and didn't care who watched.

A friend offered to package Shungnak's body so it could be sent home to Sitka. Eyes red from crying, Richard dropped the box at the air freight office. Nita called as soon as the makeshift coffin arrived. She told him about the grave site she had chosen, tucked behind the rhododendron bush in the backyard.

A few weeks later, Richard cleared his desk, turned in final grades and office keys, and boarded a flight home. Ethan and Nita met the plane—Ethan, taller, a cocky preteen; Nita, quiet and sweet as ever. At home, salmonberry bushes glowed with brilliant green. A song sparrow sang from beach grass. A sapsucker hammered the mountain ash tree in the front yard.

Richard kept glancing behind the house as he hauled gear and prepped the boat the next morning, but he couldn't muster the courage to climb the hill and visit Shungnak's grave. He and Nita shoved off midday, high clouds thinning to hints of blue. Murres, murrelets, gulls, and seals peppered a smooth sea. Mount Edgecumbe rose into a pillow of clouds, high slopes streaked with ravines of snow. Nothing unusual. Everything familiar, beautiful.

On the island, they moved slow, holding hands, talking in low tones, reveling in the peaceful still-brown muskegs and deep forest shade. They settled on the beach, nestled near a cliff, out of the wind, shoes off in the strengthening sun.

SPRING WIDENED INTO A SUMMER rich with surfing and fishing and exploring, desk work limited to a few magazine articles. Susan Bergholz called from New York with weekly updates on the book—all good news. Strong reviews, magazines wanting excerpts, radio stations scheduling interviews. September, the high slopes shifted from alpine greens to a suite of yellows and browns. October, the first snows capped the peaks in brilliant white.

The arrival of a package from the publisher was startling, although not unexpected. Richard sat alone at his desk, small box heavy in his lap, and looked over a wind-streaked sea. He opened the box slowly, held the hard-bound volume in both hands. No excitement, no elation—just a calm, deep pleasure. And then, a flood of sharp sadness.

Tucking the book under his shirt to protect it from rain, Richard walked behind the house and found Shungnak's grave, a scar of raw earth covered with rocks. He squatted and cried for his dog. He wished she could come back for just this one day, to share the opening of the gift born of Koyukuk ravens and Chief Henry's stories, of hidden deer and unseen scents, of wind in trees and rain on tents, of the candlelit shack with its scratching pencil and scurrying mice, of harrowing seas and moonlit runs, of a broken heart healed and opened by the slow patient love of place.

Richard's shoulders grew wet; his ribs shivered with cold. The rain slowed to a mist. Gratitude pushed the last tear of sadness from his eye, and Richard turned for the warm lights of home.

PART IV

True Wealth

KETA

Nita plugged her ears against the din of barking dogs. Richard paced beside her, pausing to gaze at each animal. They stopped alongside a silky-furred black-and-white border collie. She stood on her hind feet, forepaws on the wire mesh of her cage, ears back, bright-eyed, panting. She was built like a fox, with a long white-tipped tail, white around the collar, dappled socks and feet.

The kennel manager allowed them to visit with the dog in a concrete room. The border collie vibrated with a quiet intensity. She alternated between nudging and snuggling up to Richard and Nita, and looking out the one window, paws on the sill.

They told the manager they'd think about it, maybe return the next day. Richard slept poorly, gripped by the prospect of freeing the caged dog. With Nita's blessing, he returned to the kennel after breakfast. He signed the papers, paid the fee, and brought the little dog home.

In the late evening, the forest trail rich with birdsong, Richard watched the dog watch the world. He pondered names and settled on Keta, from an Iñupiaq word meaning "go ahead" and the scientific name for chum salmon, *Oncorhynchus keta*. He tried the name out loud. Keta stopped and looked back, ears perked, eyes bright.

While Richard loaded the skiff for a run to the island the next day, Keta sniffed the water's edge, then shied away from the gentle waves. Richard scooped her up and placed her in the boat. She was an ocean dog now and would learn, soon enough, to not fret wet feet.

In the middle of Sitka Sound, a muscled swell rolled under a glassy sea. Fulmars, auklets, cormorants, murres, gulls, and murrelets dove and flitted over waters ripe with the rich pulse of summer.

Richard had not been to the writing shack for nearly a year. The door had sagged a bit, and it scraped against the floor when he pushed it open. He swept mouse turds from the bunk and counter, filled Shungnak's old water bowl—half of a hard-plastic fishing buoy—and placed it by the stove for Keta.

That night, Richard journaled by candlelight, Keta curled by the door. A brilliant full moon glowed pale in the shack's single window. The gentle sound of the surf rose with the tide.

I'm slowly re-establishing myself on the Island, gradually doing more of the things I did before my long writing (then teaching) binge. This overnight at the cove is a re-habitation of the place I left to write and then couldn't come back after Shungnak died. Now, with Keta as my new partner, I can pick up where I left off.

I have that old feeling inside, the Island afterglow, that euphoria. Once again I ask myself where in the world I'd rather be, and there is no other place. Everything is exactly where it belongs; everything is perfectly in its place at this moment.

RICHARD AND NITA HELD HANDS walking the short path to Steve and Andrea's house. The sweet smell of barbequed salmon wafted from an outdoor grill. Inside, balloons and crepe paper hung from the ceiling. A half dozen close friends had gathered to celebrate the publication of *The Island Within*. Everyone in the room had been to Kruzof with Richard. Some had surfed and hunted; others had camped and beachcombed. All knew the love and hard work poured into each page. It was an evening of easy laughs, inside jokes, and tender stories. After a champagne toast, Richard choked up, expressing gratitude for his friends, tears more honest than words.

The next morning, he boarded a plane to begin his West Coast book tour. That evening, hunched in a rental car caught in frenzied I-5 traffic between Seattle and Bellingham, he missed his exit and arrived late to his first reading. *Maybe no one showed up,* he thought. *Maybe I'll just be able to go back home.* He found the bookstore crowded with attentive people.

The evening filled with sincere applause, engaged questions, and thoughtful praise. Richard finished the night primed and pumped for the weeks ahead. Next came readings in Seattle and Portland and then a flight to San Francisco for an interview with Terry Gross at *Fresh Air*.

According to his agent, the book was soaring—reviews in the major papers, a nomination for an award from the Pacific Northwest Booksellers Association, interest from television stations. But motel living muted Richard's initial elation; fast food and faster cars left him feeling anemic and tense.

Flying east, peering down at the endless checkerboard of farm fields splayed across Middle America, Richard tried to imagine carving a life in the flyover country. *I realize how out of place I'd be down there,* he thought, *like a flea on a bald head.*

A quick stop in Chicago, then on to Toronto and Boston. Finally, he was headed for New York City, the true megalopolis, which he anticipated with trepidation and fear. He eventually calmed down, but as the jet approached New York, he wrote this confession:

> *Crowds, crowds, crowds. I feel like I'm suffocating. The flight is fairly short, bumpy, with hazy views of flat, brown farmland and smatters of towns. I'm loath to get off the plane when we've reached the terminal.*

The cabdriver wove through dense traffic, honking, and delivered Richard to the Excelsior Hotel. The view from his fifteenth-floor window opened to a clot of buildings, tall and short, with jammed streets in between. Sirens and horns pressed through the brisk fall air.

After a shower, he met his agent down in the lobby. He followed Susan beneath the street into the subway's whir. They emerged at the office of *Parabola* magazine for an hourlong taped interview. Next, they scurried to an imposing glass skyscraper near the edge of Central Park. Vintage Books, an imprint of Random House, had bought the paperback rights to *The Island Within*, and Susan had arranged a meeting with the editor and publicist there. Seated in the plush office on what felt like the zillionth floor, Richard pretended to pay attention, but his mind kept drifting to the damp, rough walls of his writing shack hidden along the island's shore.

Back at the hotel, Richard called home. Nita's soft voice overwhelmed him with a desire to slip through the phone line and settle into their kitchen's familiar warmth. He called Barry Lopez next. From the book's conception through its publication, Barry had been there, an experienced guide and supportive friend. Once again, he offered strength when Richard needed it most.

> *Barry notices the stress in my voice and helps me to realize how wired I am. I enjoy these trips in some ways; but in most ways, I think they're treacherous: too much focus on self, too much city, too much illusion that things might be "important" when they're not, too much solitary travel in strange places, too much focus on mind and idea over sense and experience.*

After New York, it was on to St. Louis, Albuquerque, and Phoenix, more readings and workshops, motels and burger joints, interviews and interstates. Along the way, Susan sent a review from *Outside* magazine written by Annie Dillard, the Pulitzer Prize–winning author of *Pilgrim at Tinker Creek*. Early in the piece, Dillard makes an admission about *The Island Within*: "It's not the sort of thing I read, and the opening made me cringe: a guy and his dog go to an island off the Alaskan coast for the purpose of—oh, no!—self-discovery. How awful that this concept was a pandemic now affecting men."

But once she got into the book, Dillard found a sublime story so intelligently told, so filled with powerful imagery, that she wrote: "I thought, *He can't keep this up. He should have saved it for last.* The next chapter topped it. The following chapter topped *that*. And so on! . . . Anyone can see stuff and learn facts; it's what you make of it. His rhetorical pitch was as wild as Thoreau's on Katahdin, transporting as Shakespeare pushing art into the realms that ennoble the reader. I finished *The Island Within* out of breath. . . .

"You almost have to hold a gun at my head to make me read 'nature writing,' but I'll crawl over broken glass for Richard K. Nelson."

Such fine praise from a keen mind provided a needed lift to Richard in the tour's final days. Homeward bound, he stared out the window as his plane descended toward Seattle. He studied the coastal mountains riven with clear-cuts, slope after slope hacked into squares and scribbled with roads.

A tight connection at Sea-Tac Airport, and Richard boarded one last flight. He glimpsed quick views of the San Juan Islands before the plane ascended into a sea of gray. Exhausted yet unable to sleep, Richard stared into the clouds and tried to make sense of his whirlwind trip.

Based on the awards, the reviews, the stories from strangers who'd approached the podium after each reading, *The Island Within* was a success, a dream made real, a heartfelt effort to distill notions of respect, connection, humility, and humanity into a piece of art.

The plane rose above the clouds into a world of brilliant blue. In Richard's mind, Dillard's kind words swirled with the images of hacked mountains on the edge of a cutover continent. The only thing Richard knew for sure was his gratitude in returning home.

The plane dropped below the clouds, and Kruzof Island was suddenly, vividly there. As he drank in the high-sloped mountain, the tangled, surf-shrouded shore, the patchwork muskegs, he came to a realization:

> *What a gift to realize that the source of my happiness is so wholly attainable, not some distant illusion of achievement or fame or wealth. I have been chasing the wrong dream, focusing my fantasies on writing, which is one outgrowth of what is meaningful for me, but not the source of that meaning.*
>
> *I must return to the grace of living outdoors. And I must return to the principle that my professional career is always secondary to the conduct of my personal life and my life in nature.*

BOOTS AND BIKE LOCKS

Richard eased his skiff ashore in the thin, predawn light. A large inflatable coasted in beside him, GREENPEACE painted along the length of both pontoons. He cut the engine and stepped onto the gravel beach beneath his house. A dozen people emerged from the shadows and clustered around the boats. Most wore life jackets. Some had cameras. Everyone carried a handheld radio.

The Greenpeace captain spoke in a low voice. "Morning, everyone. Thanks for showing up. Remember, we can't predict what the mill workers will do. We can only control our own actions so let's stick to the plan. How about a radio check before we go?"

Richard shifted from foot to foot. Intentional confrontation was not his way. Since opening in 1959, the pulp mill had provided solid salaries and a generous retirement to two generations of workers. The mill's $18 million payroll represented 25 percent of Sitka's economy. Many Sitkans, including Richard's closest neighbors, believed the health of the mill and the health of town were inseparable; stop the flow of trees and Sitka would wither.

Richard had arrived in Sitka with a notion of neighborliness that transcended political, religious, and cultural differences. Getting along was more than etiquette. It was a way of life, an orientation of spirit stretching back to his kindness-at-all-costs midwestern childhood. But when southern Kruzof Island landed on the chopping block, his sense of courtesy collided with his love of place.

Kruzof was not pristine; much of the island had already been logged. Richard had explored the overgrown roads winding across the island's center. He'd forced himself to traverse a clear-cut once, years ago, slipping and stumbling through a chaotic jumble of limbs and stumps. It took over an hour to move a couple hundred yards. He clambered above the debris onto a six-foot-wide stump. He knelt on the soggy surface and, using his knife as a pointer, counted the rings. The tree had died in its 423rd year.

I stand to see the whole forest of stumps. It looks like an enormous graveyard, covered with weathered markers made from the remains of its own dead. Many of the surrounding stumps are smaller than my platform, but others are as large or larger. A gathering of ancients once stood here. Now it reminds me of a prairie in the last century, strewn with the bleached bones of buffalo.

He tried to imagine the tree's beginning: a seed dropped by a crossbill, heavy rain washing it deep into the moss. The first delicate needles would have sprouted in the years soon after Christopher Columbus arrived in North America. As a sapling, it would have stretched skyward as the first Europeans explored the Appalachian foothills. As a mature giant, its high branches would have swayed in the winds that filled the sails of Russian fur traders. In the years before it fell, it would have bent before the breezes wafting exhaust from the stacks of cruise ships.

Richard closed his eyes and conjured a raven's-eye view of a man and his dog creeping through the intact forest on the island's southern shore. He looked down on towering spruce and hemlocks, high limbs whipped by a gale. He visualized deer, hungry in the snowbound days of late winter, bedded beneath protective branches. He felt spring's warm breath and heard the rustle of dragonfly wings, the rich notes of a hermit thrush, and the chatter of a red squirrel. He then opened his eyes to the quiet, gray, devastated valley.

For a decade, Richard had attended Sitka Conservation Society's weekly meetings. He testified at public hearings, wrote letters to the editor, sent pleas to his representatives. Tedious, thankless, unglamorous work—the backbone of conservation. But when the US Forest Service scheduled another round of Kruzof logging, Richard agreed to take a turn as president of SCS.

The notion of stripping the living flesh from the island's southern forest felt akin to slowly dripping poison into the bodies of his closest friends. Resistance was now his full-time work, his twenty-four-hour focus. When Greenpeace's *Rainbow Warrior* dropped anchor in Sitka Sound, Richard joined a meeting between the crew and local activists.

The Greenpeace folks were reluctant to engage in direct action without local participation. Don Muller, the bookstore owner, was eager and ready. Don first came to Sitka to work as a chemist at the pulp mill. In his first five minutes on the job, he knew he wouldn't last. He stuck it out for two years, enough time to learn the ugly chemistry of pulp production. When he became a vocal opponent of the mill, some of his ex-coworkers grew bitter. Don tried to keep it civil. He'd wave on the street. Smile in the post office. But people didn't always respond well. One afternoon, he saw Chris Skoog, a broad-shouldered bear of a man, in the grocery store. Don stopped his cart and said hello. Chris growled and pushed past.

As required by state law, the mill produced annual emissions reports. Don read them all. He knew the cloud of sulfur rising from the stacks and dioxins seeping into the sea were far greater than reported levels. The true emissions violated both the Clean Water Act and the Clean Air Act. Don decided if the mill was breaking the law, he would too.

Anchored near the *Rainbow Warrior* was a barge carrying a railroad car full of chlorine. Don explained to the protesters how the tanker of chlorine was transferred from the barge into the mill. Built around his knowledge, they devised a plan to disrupt the chlorine delivery.

Don and another volunteer would each lock themselves to one of the dock's two big cleats. One U-shaped bike lock would go through a hole in the cleat's center. A second lock would loop through the first and go around the protester's neck. Two additional protesters would suspend themselves with climbing gear down the face of the dock, a banner strung between them.

Others volunteered to coordinate with the local paper, alert the radio station, call the police. Richard agreed to drive a boat and take photos. His house was a short ride from the mill. They'd launch from there.

Gathered in the predawn dark, Don was excited, giddy, clear about the convergence of action and conviction. Richard was nervous. He glanced at his neighbors' homes, glad to see the windows still dark. He'd worked hard to get along with these folks, to look past their ardent support for the mill. When they met in the side yard, they'd talk fishing and boats and weather, steering clear of politics. Richard worried his participation in the protest was a violation of a tender, tacit truce.

The August sky glowed with the first blush of dawn as the boats zipped into Sawmill Cove. Richard scanned the dock. No one in sight. Don fiddled with his two bike locks, suddenly anxious about having one clamped around his neck. The skiffs pulled up to the dock face, the air thick with creosote. Protesters scurried up the ladder and unfurled a twenty-foot-wide yellow banner:

ALASKA PULP CORPORATION
POISONS THE BAY
PLUNDERS THE FOREST

Already wearing climbing harnesses, two protesters quickly suspended themselves on either side of the banner. Don snapped one of his locks into a cleat. He stood for a moment, second lock in hand.

"*Hey!* What's going on?" a man shouted, hurrying across the yard from the mill. Don knelt, slipped the lock around his neck, clicked it closed.

Bob Ford, a stout man in his early forties, strode up to Don, coffee cup in hand. Don looked up, Bob's black leather boots inches from his face.

"Morning, Bob. How you doing?" Don said.

Bob chuckled. He peered over the dock and saw the banner and dangling protesters and threw back his head with a full-throated laugh.

"How the hell did you do it? I've been here all night. Just went for a cup of coffee. I wasn't gone but ten minutes."

With the *Rainbow Warrior* in town, the mill had anticipated trouble and had assigned Bob as night watchman. What appeared to be stealth on the part of the protesters was nothing more than blind luck.

Chris Skoog, the grocery-store growler, was next on scene and, unlike Bob, he wasn't laughing. He grabbed a pike pole and began pulling and tearing at the banner. The morning shift trickled in. The dock filled with more boots and a darkening profanity. Don tensed, on his hands and knees, vulnerable to the stirring mob. Only when Ed Green, the local cop, arrived did Don relax.

When bolt cutters failed to crack the bike lock, workers hauled an electric pipe saw from the mill. The whine of steel on steel rang loud in Don's ears. Metal flecks peppered his cheeks. Lock severed, Don jumped to his feet, glad to rise above all those boots. Officer Green cuffed him and placed him in the safety of the squad car.

Don was booked and released before lunch. He returned to the bookstore the following morning. That afternoon, twenty mill workers gathered on the steps of Don's store. Most carried signs.

HEY GREENPEACE!
HUMANS LIVE
ON THE PLANET TOO!

SHUT DOWN!
TODAY THE MILL
TOMORROW THE CITY OF SITKA

GREENPEACE EQUALS
ECONOMIC TERRORISM

After a half hour of pacing and some half-hearted chants, Russ Ferguson chained himself to the store's front door. Russ worked at the mill. His wife, Nancy, worked for Don. Small towns weave tight webs. You're caught the moment you arrive.

Russ wore an orange T-shirt. The front read SUPPORT THE TIMBER INDUSTRY: BUY BOOKS HERE. The back of his shirt read CAUTION—CHLORINE BLEACHED PAPER SOLD HERE.

Don admired a man willing to become a convict for his convictions. Rather than call the cops, Don served cookies and coffee. But come closing time, he couldn't close. Russ's body was in the way. One more round of coffee did the trick. Around nine o'clock, Russ dug a key from his pocket and unwound the padlock and chain. He mumbled something about being done with this nonsense and scurried off to empty what had to have been a painfully full bladder.

As Don served cookies to the protesters, Richard was at home rereading *Civil Disobedience*. Henry Thoreau was appalled to be a citizen in a country where slavery was legal. "What I have to do," Thoreau wrote, "is to see, at any rate, that I do not lend myself to the wrong which I condemn." One hundred forty years after Thoreau published, Richard explored similar ideas in *The Island Within*. With a stump as a witness stand, he placed judgement on himself.

> *I hold few convictions so deeply as my belief that a profound transgression was committed here, by devastating an entire forest rather than taking from it selectively and in moderation. Yet whatever judgement I might make against those who cut it down I must make against myself. I belong to the same nation, speak the same language, vote in the same elections, share many of the same values, avail myself of the same technology, and owe much of my existence to the same vast system of global exchange. There is no refuge in blaming only the loggers or their industry that consigned this forest to them. The entire society—one in which I take active membership—holds responsibility for laying this valley bare.*

"Let your life be a counter-friction to stop the machine," advised Thoreau.

Don's neck in the bike lock applied direct friction to the machine. Others in the conservation movement added pressure through political proceedings and courtroom lawsuits. But Richard didn't share Don's tolerance for conflict. He lacked the fortitude and patience to push through the dense language of bureaucracy. He was, at heart, an artist. And like a thrush defending its territory with song, he'd do his part by defending the forest with words. He'd spin society a better story, tell a deeper truth, pose a better question. He'd fight lies with lyricism, greed with gratitude. He'd blur boundaries by exposing the truth of our connections.

FORGOTTEN LANGUAGE

Richard often returned to the broken bone. Sometimes he brought a friend; mostly he went alone. He'd find the misshapen fused deer leg and ponder the will to persist, the force to become whole after being shattered.

His final visit to the bone was not planned. While hunting with Keta, he found himself in the muskeg sloping toward the hemlock where he'd sat with Gary Snyder all those years ago. He leaned his rifle against the familiar tree and combed fingers through moss, feeling for the hard lump. He brushed off the hemlock needles and dirt and ran his thumb over the pointed shard that had poked through the deer's hide. He imagined the years of pain, the steady throb and ache.

When he turned to go, he slipped the bone in his pocket. At home, he placed the healed leg above his desk.

The year after publication of *The Island Within*, Richard was torn between an impulse to begin a fresh project and a desire to never write again.

> *Sometimes I wonder why I have this urge to write, to spend months imprisoned at my desk, inside the house, when I could be working outdoors, using my whole body, doing activities that involve no stress at all. Or when I could make a good income teaching just four months of the year, then spend the rest of the year entirely free. What is the madness that drives me to write? To spend months in solitary confinement, plagued by insomnia and all sorts of nervous symptoms, working on books that bring modest recognition at best*

and far too little income? I can say, at least half seriously, that writing is a sickness.

The book had just been awarded the John Burroughs Medal and translated into Japanese. Susan Bergholz, his agent, assured him his next book would be a financial and literary success. But Richard promised himself, and Nita too, he'd never again endure the stress of poverty. Unless his next book was supported by a generous fellowship or healthy advance, he vowed to overcome his writing addiction.

But there was no harm in dreaming.

He contemplated a book about Low Island, a two-acre shoal off Kruzof's southern shore. He fantasized about the intimacy possible in such a tight geography—recognizing individual hermit crabs in a tide pool, distinguishing one harbor seal from another, delving into the land by diving deeper into the details. He also mused about a book on deer, an opportunity to once again be on Kruzof Island and with the animals he so loved.

Susan urged him to expand the scope, avoid a reworking of the island book, enlarge the readership with a wider geography. So Richard expanded his focus from Kruzof deer to North American deer. His proposal for *Heart and Blood: Living with Deer in America* generated strong interest from the New York houses. That August, Susan called with the news of a generous bid from Knopf—half up front, half upon delivery of the finished manuscript.

Suddenly everything looks different to me. I feel a great oppressive burden lifting . . . after these months and months of fretting over what to do next. I'm surprised by the intensity of my reaction. Partly it's because I will not have to leave home to teach. But I suspect the deepest happiness is because this affirms my work.

I find myself constantly thinking about deer and the project. I'm running words and phrases and descriptions through my mind. The momentum is building.

Richard had two years: twelve months for research, twelve for writing.

He made a series of calls that set a pattern for his inquiries. Each conversation ended with a list of names of more people eager to talk about deer. With their natural predators exterminated, deer populations across the continent had surged, becoming a growing bane for farmers, hazard for drivers, pest for gardeners, and target for hunters. The peaceful deer was the catalyst of controversy all over the country.

Deer tracks led Richard from Texas to New York. He shadowed hunters and rode along with ranchers. It was an anthropological romp through the rural counties of his own culture. He dug deep into the deer's story—perhaps too deep. One year stretched into two, each encounter opening into a fresh opportunity. Three years after signing his book contract, Richard had dozens of journals filled with stories and a file cabinet packed with research papers but had yet to write the first chapter. Fifteen percent of the book's advance had gone to his agent. The remaining money, stretched across years of travel, meant he was, once again, living and writing in poverty.

With a felt-tipped pen and a small board, Richard made a sign:

WRITING

PLEASE DO NOT DISTURB

He tied a bit of cord through holes drilled in the upper corners and hung it on his office door.

Months of isolation crushed his spirit. Hours of sitting wracked his body. A pinched neck nerve flashed pain down his left arm. Each morning, he donned a neck brace connected to counterweights suspended from the ceiling over his desk and put himself in traction.

Multiple times a day, he'd lift his fingers from the keyboard, gaze longingly out the window, then refocus on that shattered deer leg above his desk. If that deer could limp through years of agony, he'd see this book to its end.

Heart and Blood is a comprehensive, meticulous, eloquent exploration of our collective relationship to the land. Interspersed with journalistic chapters are lyrical firsthand encounters with deer. It is a testimony to Richard's artistry that the beauty of the prose does not reveal the ugliness of his writing process.

The book would go on to earn the Sigurd F. Olson Nature Writing Award, and would solidify his selection as Alaska State Writer Laureate. In the final passage, he shares the realization of a long-held dream:

I've now seen a wild deer born, and on this wild island where my love for deer was born. I remember what the Koyukon elders teach: that everything we receive from nature comes to us as a gift.

The fawn and I live from these same earthly gifts—the air we breathe, the water we drink, the food we eat. Looking at the fawn, I see myself, being born and flinging out into the world, to live and grow and die, and someday to feed other life, nurturing further generations in turn. Because I hunt in these muskegs every fall, our fates might someday conjoin. For this I feel neither guilt nor sadness, only gratitude and joyful affinity.

Lovely deer, you are always in my heart, dancing down the dawn into the light. Lovely deer, you are always in my blood, dancing down the dusk into the night.

When Richard was almost done writing, his friend Jim Dodge sent along a copy of one of his poems. Richard tacked it to his office door.

SILENT NIGHT
I would rather crawl naked over the Himalayas
With a leaky 55-gallon drum of radioactive turds
Strapped to my back
And a bullet through a lung
Than sit down alone in a room full of language
To write another word

It was still there two decades after he had finished his last book.

THE FORGOTTEN LANGUAGE TOUR WAS born from Marion Gilliam's love of the theatrical arts and appreciation for literature. As publisher of the environmental magazine *Orion*, Marion was in contact with the country's finest nature writers. He envisioned infusing the passion of the written word with the vitality and immediacy of public performance. He imagined a band of troubadours traveling through the modern United States, stopping in towns large and small, giving voice to the life being buried beneath the pavement and malls.

Marion drew inspiration for the name of the tour from a poem by W. S. Merwin:

> WITNESS
> *I want to tell what the forests*
> *were like*
> *I will have to speak*
> *in a forgotten language*

The poem embodied Richard's life. In the three decades since mushing and camping with hunters on the frozen Chukchi, Richard had studied ice and wind, seals and caribou, dogs and moose. He'd joked in Iñupiaq and prayed in Koyukon. He'd absorbed the songs of birds and studied the tracks of deer.

In wild solitude, he took Tagruk's lessons and Catherine's stories and made them his own. With the heart of a hunter and eye of an artist, he refined the language of connection and put it on the page. And with the urge of an activist, he accepted Marion's invitation to the Forgotten Language Tour and went on the road.

In a caravan of cars, a shifting cast of characters traversed thirty-seven states over the course of three years. Writers came and went, joining the rolling road show for a few weeks before returning to their private lives.

Richard had resisted touring for *The Island Within*; he was a writer, not a showman. He grew weary of his own words repeated night after night. The Forgotten Language Tour was different. It replaced the one-man show with an amalgam of the United States' keenest artists and finest thinkers: Alison Deming, Pattiann Rogers, Robert Pyle, Rick Bass, Scott Sanders, Janisse Ray, Terry Tempest Williams, and Barry Lopez. Richard felt awed and honored to be included.

At each venue, Marion dimmed the lights and set the mood. Passages and poems appeared one after the other. Each night brewed its own brand of magic, a living piece of art that, like a sculpture of ice, would melt at evening's end and be created anew in the next town.

"I don't believe any audience member ever left without being transformed," Marion recalled, "without, for a moment, experiencing joy. And,

I must say, Richard's artistry formed the center of that joy. His presence disarmed any whiff of egoism between the writers. His stories from the runners of a dogsled captivated and transported. He was truly a bard bearing gifts from a foreign world."

The writers traveled together, ate together, brushed their teeth together. They laced up for morning runs, stayed up for late-night talks. Gary Nabhan, an Arizona-based ethnobotanist, says Richard held the center of a strange blend of mysticism and mischief. "He's so on target," he said, "with his moral, poetic, and ethical voice. And, at the same time, he's got this intense playfulness. His humor is ferocious, rooted in his love of place. He cares so deeply about the world and its cultures, the animals and landscapes, that his heart is constantly being broken. Humor is the counterbalance. Without it, he'd be dead by now."

At night, they'd speak of lost species and dying cultures. The next day, they'd get kicked off a miniature golf course for having too much fun. "We were a gang of merry pranksters," Gary Nabhan recalled. "We ruined places with our rowdiness. We were ravens and coyotes in a culture of pigeons and mice."

Across the country, Richard met people craving connection, aching, somehow, to become rooted in America's fragmented landscape. He leaned into the creative challenge, using story to assuage the collective hunger. And Richard was nourished in return. Cloistered within the gathered crowds and circle of writers, he dared to believe in the possibility of change.

Richard became the backbone of the tour, performing in more events than any other writer. He augmented the tour's small stipend with additional paid speaking gigs at conferences and college campuses. His yearning for Nita, Keta, and the island was eased by his commitment to the work.

Don locked his neck to a pulp-mill cleat. Richard went on the road to bring life to a forgotten language.

In the summer of 1993, Richard was home between trips when Nita dropped the morning paper on the kitchen table. He stopped chewing midbite, set his cereal bowl down, and read the headline twice.

ALASKA PULP CORPORATION SAYS SITKA MILL TO CLOSE

Shifting economics and growing public opposition had finally pressured the federal government to stop the promised fifty-year supply of trees. Richard jumped on his bike and rode downtown. Don was already at the bookstore. Others soon trickled in, their elation tempered by disbelief. After decades of resistance, it was hard to accept Sitka would soon be free of the acrid-smelling orange smoke and dense tea-colored effluent, difficult to trust the trees they loved would really be left alone.

They celebrated privately: no honking horns, no jubilant shouts. They knew their relief hinged on the stress of neighbors now looking for work. Richard returned home, laced his running shoes, whistled for Keta. Jogging the forest trail, he thought of all the meetings, public testimony, lawsuits, and protests, all those seemingly fruitless hours now bearing fruit.

Near the river mouth, he stepped to the beach and gazed across to the island. Relief swelled into tears. It was like having his closest friends cured of cancer or saved from disfigurement, the avoidance of heartbreak he wasn't sure he could survive.

The Forgotten Language Tour gave rise to a gathering on the banks of the Potomac River in June of 1999. The event, named Fire & Grit, drew in hundreds of people. For three days, folks mingled in the shade of enormous tents. Evenings, they packed into an indoor auditorium to hear the country's greatest writers. Richard was the last. He took to the podium in a packed house—every chair filled, people leaning against the back wall, seated on the stairs. So many faces of people he'd come to love—people who had been to the island, who understood his passion and tender heart.

He adjusted the microphone and began with a simple story: a walk in the Alaskan woods with a close friend and a sudden encounter with a brown bear in a dense thicket of alder. Richard described how the bear heaved up onto its pillar legs, listed from side to side, then disappeared amid the clouds of leaves and ferns.

"The bushes shook, we heard footsteps pounding heavily away, and then came a protracted, reassuring silence," he said. "Through the rest of the day, I couldn't stop thinking about our incredible good fortune—to encounter a truly wild grizzly bear in a truly wild land, at a time when both have become heartbreakingly scarce. To be in a place where nothing at all has vanished from the natural community, where you do not feel the vast, oppressive emptiness of extinction."

Richard then pivoted to his college years and his participation in marches and protests against the Vietnam War and the denial of civil rights. He confessed that, like so many of his generation, the notion of patriotism did not rest easily in his soul. He shared how, in recent years, he'd come to view the American flag in a new light.

"A patriot," Richard told the audience, "is one who loves his country and zealously guards its welfare. We are creating, as did Native Americans long before us, a patriotism based on ecological knowledge, moral consideration, ethical principle, spiritual belief, and a profound love for the earth underfoot. I believe this is the most basic, most urgent, and most vital patriotism of all, because conservationists are working in service to the elemental roots of their existence, as human organisms, as members of their communities, and as citizens of their nation's land.

"With my own eyes, I can see the results of our work: whole mountainsides, broad valleys, and sprawling islands covered with lush, living forest. Places that would have become barren lands of stumps and slash if we had not used our voices.

"I am expressing my growing sense of allegiance to this living nation.

"Allegiance to the forests, the prairies, the deserts, the mountains, the swamplands, the seacoasts, the lakes and rivers and oceans.

"Allegiance to the gardens, the wheat fields, the dairy farms, and the rangelands, from which our bodies are made each day.

"Allegiance to the greater community of landscape and organisms that encompasses and sustains us all."

Peter Forbes, a leader in the land conservation movement, was in the auditorium that evening. When Richard finished, Peter rose to his feet with everyone else. "His words were powerful," Peter recalled. "But what I remember most was his smile, his exuberance. I wanted to feel that alive."

HOME, LAND, SECURITY

When I first read *The Island Within*, I thought: *Now here's a guy telling my story.* I hunted and explored the same coastal forest. I shared his obsession with wild country. I understood his disillusionment with greed and destruction masquerading as the American dream.

I underlined beautiful phrases and inspiring passages, including a description of Koyukon elders saying each type of weather has its own spirit and consciousness.

> *If this is true, there must be a spirit within every raindrop, as in all else that inhabits the earth. In this sense, we are two equal forms of being who stand in mutual regard. I bend down to look at a crystal droplet hanging from a hemlock needle and know my own image is trapped inside. It's humbling to think of myself this way. In the broader perspective of earth, I am nothing more than a face in a raindrop.*

A year after reading his book, I met Richard on the shore of a forested lake. He'd hiked in from the south with a cadre of Sitka friends. I'd trekked from the north with a few Juneau buddies. The teams met with the goal of mapping some of the largest trees remaining in the Tongass. We pitched camp on damp moss and gathered around a hissing, smoky fire.

I'd expected a poet—a contemplative man moving slow among the trees, capturing lyrical lines in a notebook stashed in the breast pocket of his wool coat; I'd envisioned some sort of monk (who else sees himself as a face in a raindrop?). I encountered an enthusiastic, quick-witted man dropping f-bombs like a toddler spills Cheerios.

I don't remember who threw the first spruce cone, but for three days, while the rest of the crew measured and mapped, we ambushed each other, pockets packed with bristling bullets. We chased and scrambled, scurried and laughed. It didn't matter that I was twenty-three, Richard just shy of fifty—we were both kids loose in a forested playground.

Richard was a literary celebrity, an iconic Alaskan voice. I was, at first, a bit awed by his presence. But rolling around in the wet moss and deer turds, we built the foundation of a lasting friendship.

Richard had just published *Heart and Blood*. I had just built myself a home. He'd sworn off writing and yearned for the island. I worked in an office and ached for the woods. We broke that first camp on a drizzly morning. As we shrugged into our packs and went our separate ways, I whizzed a departing spruce cone past his ear.

Weeks later, a copy of *Heart and Blood* arrived in the mail with an inscription: *Here's hoping for the chance to swap stories over many more wet-wood fires.*

We did, in the following years, find ourselves feeding damp sticks into crackling flames. We shared a passion for drumming and spent hours beating on hollow logs and battered buckets. Richard crafted didgeridoos from washed-up plastic pipe; we'd play and laugh until our arms and lungs gave way, then collapse into a tent and rest up for the next day's adventures.

In those days, Richard was on the road more than he was in Sitka. Leaving home to talk about his sense of place was, after years of touring, wearing thin. His appetite for travel ended abruptly in the fall of 2001. The Forgotten Language Tour had just finished an East Coast run. He was in the Boston airport on the eleventh of September. Billowing smoke and anguished faces flashed across television monitors. SWAT teams armed with assault rifles swarmed the terminal.

When air travel resumed, Richard sat on a half-filled plane, his mind consumed by collapsing towers and rising rhetoric. The analysts and news anchors talked about how 9/11 had changed the world.

Richard wasn't so sure.

Pearl Harbor was bombed the week he was born. Sixty years later, black plumes rose from the south end of Manhattan. What, if anything, had truly changed in the span of time and space separating those attacks?

The Caribbean island where Columbus put ashore on a windy Friday morning, has, in the intervening centuries, been essentially scraped clean of all indigenous life. The incursion that began on the most seaward side of the Bahamas continues with the removal of mountaintops in Kentucky, the stripping of forests in the Cascades, and the search for oil in the Chukchi Sea.

The sudden aggression inflicted upon Hawaii and New York was brutal and tragic but no more so than the ongoing desecration of North America by our own hands.

Home.

Land.

Security.

Who really knows about such things? The pundits? The politicians?

Richard peered out the plane window and watched Midwest cornfields give way to West Coast clear-cuts. He thought of the few folks he knew who were truly secure and at home on the land.

Tagruk Bodfish, who, peering into mist-shrouded waters, asked in hushed tones for a seal to come. Catherine Attla, who prefaced answers to questions by saying, "What little I know about that . . ." Chief Henry, who, on his deathbed, expressed gratitude for all the nights camped beneath a spruce tree roasting grouse over a fire.

In contrast to the armed intensity of the Boston airport, the Seattle terminal was hushed, subdued. On the final flight north, Richard felt little beyond bone-deep exhaustion. When the plane descended from the clouds and Kruzof Island popped into sharp, almost touchable relief, he knew he was done.

No more travel.

It was time, once again, to let his feet sink into the thick moss.

Keta was, as always, delighted to be reunited. She pranced on the beach as Richard loaded the boat.

On the island, Richard dropped his gear in the writing shack, then slipped into the forest. He paused on the edge of a still muskeg. Treetops tore gray strands from low clouds. A squirrel chattered from the shadowed woods. A raven huffed past, called once, then again. A light mist whispered into the surface of a pond. The world of suicidal hijackers faded, breath by breath, beneath the gathered silence.

Back at the shack, he kindled a fire, hung his damp coat by the stove. Eating dinner by candlelight, Keta curled by the door, Richard could almost believe all was right with the world. But that night, he could not sleep. He listened to the dripping rain and scratching mice and thought about the changes in the country and in his own home.

Ethan had finished high school and was working as a mechanic in Portland. Nita, too, was ready for a change. Her decision to move out was, unlike Kathy's sudden departure, not a surprise. Richard's marriage to the wilds and devotion to his work had long been a burr in their relationship. Years of travel had proved too much. Once-warm intimacy had faded beyond reach.

Helping Nita pack her things was heartbreaking but not as traumatic as Kathy's single-sided decision. Loneliness is hard enough without being laced with the pain of betrayal. His empty bed and quiet kitchen were, at times, more than Richard could bear. But his circle of friends filled his home with warm food and welcome laughter.

He also carried the island in his bones. A simple glance at Kruzof's familiar shape softened the edge of grief. The island would never leave. The shack, the deer, the storms, the surf, the raven, the trails worn by boots, paws, and hooves—it was all there, all the time, just a boat ride away. Even on the darkest days, he knew he was not alone in this world.

BEST DAYS

Richard was sixty and on his own. He had no job, no retirement fund. He did have a self-imposed ban on travel outside of Alaska and a fierce determination to be free of his desk.

The start of his next career literally washed up on the beach.

For a decade, Sitka fisherman Eric Jordan had done live in-studio interviews for a local radio program called *Fin & Feather*. Richard had, through the years, filled in when Eric was out of town. When Eric was ready to retire, Ken Fate, KCAW's station manager, asked Richard if he would take it over.

"Only if I can do it outside," Richard said. "I don't want to be trapped in a studio."

"That's possible," Ken said. "It's called live-to-tape. There are no rules. That's the beauty of a small station. We can try anything."

Microphone in hand, digital recorder in his pocket, Richard approached a beached and bloated whale. A brown bear clawed and chewed at the rancid carcass. Richard crept close. With an adrenaline-laced whisper, he described the bear tearing chunks of flesh, the whale's blubber rippling with each powerful tug. Richard recorded for thirty minutes, lacing natural history with direct observation. Ken knew that first program was a gem. Listeners agreed.

For his next show, Richard strapped himself into the high branches of a hemlock tree exposed to the full fury of an Alaskan gale. Buffeted and battered, he raved about the storm's power and described the wind's spirit as perceived by his Koyukon teachers.

He shimmied out of that wind-whipped tree, excited about the new work. Radio had it all: the crafting of stories without a desk, an audience without having to travel, an opportunity to celebrate wild beauty without leaving home, and, most importantly, the chance to keep learning and listening. Richard and Ken called the new program *Encounters*.

Hidden beneath the casual, spontaneous tone of each show were weeks of research and planning, preparation and travel. Richard mixed howling wolves, chortling cranes, snorting musk ox into his own narrative with a fluid precision. "It was a pleasure watching him work," Ken recalled. "He got so good so quickly."

IT WAS LATE SUMMER 2012 when Richard called me to say he was heading to Glacier Bay. "I'm prepping three programs: black bears, mountain goats, and tidewater glaciers. Glacier Bay is the one place I might be able to record them all."

"I'm in. I'll have the boat ready," I said.

He arrived in mid-September. We loaded gear and pushed off on a chilly morning. As we cruised toward the ice, gray clouds thinned into a rare stretch of fall sun. Late afternoon, we nosed onto a beach and made camp in the treeless, deglaciated terrain. Glistening bergs drifted past on the ebbing tide. Soaring peaks bristled with fresh snow. We'd not seen another boat all day. I anchored the boat and paddled back to shore. Richard unpacked his recording gear.

"Put these on," Richard said, handing me a pair of headphones. He then gave me his parabolic dish—an unwieldy salad-bowl-shaped thing with a foam handle mounted on the back. He slung a recorder around my neck and pointed to a button. "Hit this to listen. It'll change your life."

I stepped down the beach, pushed the button.

The gravel under my boots became a rockslide in my inner ear. Two-inch waves lapping against shore exploded like giant surf. A gull drifted over the cove, clucked once, and filled my skull with sound. I eased forward, slow and gentle, aware of each grinding, shuffling step. That giant bionic ear transformed the afternoon into an expansive auditory hallucination—cackling ptarmigan, yodeling loon, hissing barnacle, slippery pop of seaweed underfoot.

It was early evening when the batteries died. I slipped off the headphones and stood in a world I'd thought I knew. It was like being gifted the miracle of glasses after decades of lousy eyesight. Amplified through an exquisite microphone, vague sounds rang crisp, hushed whispers washed clean. How was it that I'd been in Alaska my entire life and never paused to fully listen?

Walking back to camp, my senses remained tuned to rustling leaves and whistling wings. My ears were not broken, I realized, just clogged with a lifetime of neglect. I felt regret for all I'd been missing, but elation for all I might yet hear.

Richard was staking the tent when I returned.

"Well, how did it go?" he asked.

"Acoustic LSD," I replied. "I think I'm addicted."

We awoke the next morning to frost on the tent and ice in the water bottles. Richard stomped his feet and rubbed his hands, waiting for the cocoa water to boil. "Freeze-ass cold up here by the glaciers. If I'd kept my focus on snakes, I'd be baking my balls in the desert right now."

After breakfast, we skiffed to the base of a mountain where, the day before, we'd seen a group of goats. We warmed as we climbed, stripping layers as we gained ground. Near the summit, we spotted goats basking in the afternoon sun. I held back while Richard, microphone in hand, eased close.

An hour later, he woke me from a sun-soaked nap.

"My God, what a piece of luck!" Richard exclaimed.

"Goats cooperate?"

"Cooperate? Holy shit! You can't believe it. The main group held still, and then, halfway through the program, a young billy walked around the corner and damn near bumped into me. Sunshine. No wind. Wild goats. It couldn't be any better."

The next morning, we boated farther up the bay. Hanging glaciers clung to bedrock walls, craggy white on smooth gray. The sea, stained a light brown by glacial silt, was jammed with bergs. At the head of the fjord, five miles away, loomed a massive wall of fractured ice over a mile across. The glacier wound upvalley and disappeared into mountains biting thirteen thousand feet into a perfect blue sky.

Richard paddling toward John Hopkins Glacier to record a radio program (Photo by Kim Heacox)

We edged closer, weaving through bergs. Sharp pops and thunderous booms echoed across the inlet. A quarter mile from the ice face, my twenty-three-foot boat, so substantial and seaworthy while tied to the dock, now felt crushable, tiny. I shut off the engine. The bergs sizzled and snapped. The glacier cracked, groaned, and then calved a massive column, with a tsunami of spray and an avalanche of noise.

Richard always recorded as close to his subjects as he could. The glacier, towering just a quarter mile from our boat, was not close enough. Richard eased the little plastic kayak off the side of my skiff. He lowered himself into the cockpit. I handed down the recording gear. He stroked toward the glacier and, in minutes, was swallowed by the maze of jostling ice.

A few hours and a dozen monstrous calvings later, I glimpsed the rhythmic flash of a kayak paddle working back to the skiff. A hundred yards away, he set down the paddle, smiled, and hollered, "*This is the best fucking day of my life!*"

He soon climbed out of the kayak, babbling with breathless excitement. "Holy balls! Did you see that last calving? Can you believe this place? There are seals everywhere up there, right at the ice edge. Oh sweet Jesus, what a day! The kittiwakes, jeez, they're feeding right there around the calving face! Can you believe this place?"

Given the wild beauty, solitude, fall sun, and his close encounter with cascading ice, it made perfect sense to me that this was the best day of his life. And it made perfect sense when he'd made the same pronouncement the previous day, after sneaking close to the goats.

What didn't make sense was my own response. Sure, I was having a good time. Sure, I was happy to be there. But the best day of my life? I couldn't match his enthusiasm. What was holding me back?

If glorious mountains, tumbling ice, and brilliant blue were recipe enough for Richard, what additional ingredients did I need? How do any of us keep our best days opening before us rather than lost in the fading folds of memories?

Back at camp, Richard buzzed with the day's beauty. "I'm damn-near seventy years old and spending time in country every bit as wild as I explored in my twenties. Can't say how good that feels."

LIKE WRITING, THERE WAS NO money in radio. Richard had to hustle for funds up front, relying on donations to cover the high cost of Alaska travel and a poverty-level salary. After a decade of producing *Encounters*, attracting nearly a million listeners, Richard would walk away.

His last show featured his most compelling recordings: bellowing musk ox, roaring wings, bugling elk, tinkling ice. With a background of howling wolves, he concluded: "The value of wildness and beauty on such a scale is impossible to calculate. It's the most rapidly diminishing resource on Earth. When I hear the choruses of wild voices, I dream of children a century from now hearing the same sounds in a world that has grown wiser because it has learned to listen."

By the time the last show aired, my own parabolic was cracked and worn from heavy use. Together, Richard and I landed a contract with the National Park Service to create a library of natural sounds. A paid gig to record the voices of Glacier Bay: oystercatchers and kinglets, rain on ponds and wind through grass, puffing porpoises and splashing fish. We pitched our tent in the quietest, most animal-rich spots we could find within Glacier Bay National Park's 3.3 million acres.

With Richard's help, I teased individual voices from the complex chorus of birds.

"That one is a yellow warbler," Richard said, pointing to a small bird singing somewhere in a thicket of alder. "It's a little higher and a touch sweeter than a yellow-rumped warbler. And over there, that trill, that's an orange-crowned warbler. It's a lot like a junco, but the pitch drops just a bit at the end of each phrase."

After an afternoon recording a bubble-blowing harbor seal and a chirping pair of ancient murrelets, we ate dinner swatting at clouds of no-see-ums. "I gotta say there's nothing I'd rather be doing," Richard said. "No pressure to produce shows, no stress of village living. Pure unfiltered listening—it's a dream."

For two years, I listened to stories rise up with the sparks and smoke of our shared fires. Stories of dogs, thin ice, and slender caribou; the ache for home, the pain of shattered dreams, new teachers revealing old ways of seeing.

"You've got to write this stuff down," I said. "It'll disappear when you're gone."

"You're probably right." Richard tossed another stick on the flames. "But I'll never do it. I'm done writing."

"I'm not," I said. "Let me do it."

NIGLIK'S RETURN

All morning, we hauled dusty boxes from the cramped confines of Richard's chilly attic. That afternoon, we sat in his small living room, journals, photos, and reports cluttered about our feet.

"Here you go. Every letter I ever wrote my parents." Richard handed me a cardboard box sagging with age. I flipped the lid open and lifted out a bundle of envelopes. The worn rubber band broke with the first touch.

"Mom was a fanatical pack rat," Richard explained. "She never threw anything away." I flicked through the envelopes, each addressed to Ma and Pa Nelson.

"Wow! Smell this," Richard exclaimed, his nose buried in another box. I set the letters aside, reached for the package. I recoiled at the stench.

Richard laughed. "Seal oil," he said. "That's what my Wainwright house smelled like all the time." Inside the smelly box were reel-to-reel tapes, each housed in a slim cardboard case. I gingerly lifted one out and gave it another cautious sniff.

"Tenacious stuff. It's not like I dripped oil on those things. They soaked up the smell just from being in my house, and they've held on to it for nearly fifty years."

"You ever listen to these tapes?"

"Nope. I imagine they were last played by my parents as soon as they got 'em in the mail."

"How about the letters? Read any of them?"

"No. Haven't touched them."

"Really? Why not?"

Richard gazed out the window framing the placid waters of Jamestown Bay. A cluster of gulls grappled over a ball of fish.

"I've never felt I had the strength. I treasure my Wainwright memories. But they haunt me at the same time. I can't explain it."

"We should go back," I said. "Let's go next fall. Let's look for caribou, walk on the—"

"Forget it," Richard snapped. "It's not happening."

When he next spoke, his voice was soft, his eyes wet with emotion.

"I can't get around my romantic attachment to what the village was. It was a magic time—still remote, the technology simple, lifeways intact. There was a tender bubble of innocence that has long since popped."

I waited, the box of tapes heavy on my lap.

Richard looked out at the gulls before speaking again. "I worry the experience of what the village has become will destroy the memory of what it once was. People's faces, Iñupiaq words, images of sea ice—they drift through my mind every day. Those memories enrich my life."

I let it go, didn't push the idea any further. We shuffled and organized and repacked. After dinner, we launched kayaks and paddled into the open reach of Sitka Sound. A gentle swell rolled under the boats. Mount Edgecumbe rose high in the western sky, tinged gold by the low sun. We drifted apart, each lost in our own thoughts.

"Check this out," Richard hollered. He'd stopped paddling and was staring into the depths. When I drew close, I saw a cloud of jellyfish pulsing beneath our boats. Each jelly was a few inches wide with short tentacles sweeping from the edge of its translucent bell-shaped body. The school of jellies spilled into the depths and shimmered beyond sight. The longer we watched, the more we saw, uncountable thousands, like breathing stars in a moonless sky.

"Have you ever seen anything more beautiful?" Richard whispered.

An evening breeze stirred the mirrored sea. We left the jellies and stroked back toward the house. Pillows of amber clouds floated past rugged peaks ablaze in the setting sun. We paddled in silence.

In the following months, I read the letters to Richard's parents and listened to the tapes, and gained insight into the trauma, loneliness, joy, and discovery underlying his first winter in Wainwright. I placed calls to the village and eventually got in touch with Griffin Lou Oenik, granddaughter of Kusiq Bodfish, the great hunter who'd first stopped by Richard's house to inspect his folding kayak. Griffin told me that Kusiq had been dead for many years but that her father, Tagruk, was still living.

"Tagruk would love to see his old friend," she said. "We all would."

When I told Richard that Tagruk was alive and eager to see him, he relented. I bought tickets before he could change his mind.

In preparation for the trip, Richard spent hours peering through a magnifying glass at contact sheets of old black-and-white photos. He selected a dozen images to get enlarged. Men butchering walrus. Kusiq scanning the sea ice from the peak of a pressure ridge. Tagruk straddling a kayak at the edge of an open lead. Hunters brewing tea in the lee of an upturned sled, dogs curled against the wind.

We packed the images along with our cold-weather gear and traveled north in the last days of October. On the long flight from Anchorage to Utqiagvik, Richard flipped through his Iñupiaq dictionary. He whispered each word out loud, making the old language fresh on his tongue.

We overnighted in Utqiagvik and hauled our bags through the snowy streets for the morning flight to Wainwright. We sat in scooped-plastic seats in the one-room airport terminal. *SpongeBob* cartoons spilled from a large-screen TV.

"This is about what I expected," Richard mumbled.

It was nine in the morning but still pitch-dark when we boarded the twin-engine Cessna. The pilot flipped on an automated recording about safety features, how our seat cushions could be used in case of an emergency.

"You get a safety briefing when you made this flight in 1964?" I asked.

"I'm not even sure that plane had seat belts," Richard said.

We landed an hour before sunrise. The pilot cut the engine. Two Iñupiaq families had shared our flight, their small children asleep. Richard and I hunched through the low door and stepped onto the snowy tarmac. Three pickup trucks eased close, headlights cutting sharp cones through the black. Two of the trucks were there to pick up our fellow passengers. They placed the groggy kids in the cab, pitched luggage in the back, and drove away. The third truck idled nearby. We approached and asked for a ride. A young white guy wearing a grease-stained hoodie rolled down the window.

"Sure. Just waiting for a piece of freight," he said. "Hop in. Cold out there."

On the short drive to town, Richard peppered our driver with questions. Originally from Michigan, Dave had been operating heavy equipment in Alaska for several years. He was in Wainwright as part of a skeleton winter crew under contract with the oil company Shell. He stopped outside a ramshackle building sheafed with blue plywood.

"There you go," he said. "The Olgoonik Hotel."

We grabbed our bags. Dave waved and pulled away. In the pale predawn gray, we could just make out the street sloping toward a dark smudge of the Chukchi Sea. No dogs, no voices, no birds. Cold stung the inside of my nose and nibbled at my fingertips.

Richard dropped his duffel bag in the snow and turned a slow circle. Across the street from the blue hotel stood a single-story gray structure bearing the sign OLGOONIK STORE. Uphill was a fire station with three garage doors aligned along a freshly plowed asphalt apron. Downslope, a string of identical homes toed the street like a regiment of soldiers.

"Honestly, Hank, I don't recognize a damn thing!" said Richard.

He settled his gaze on the distant sea. After a long pause, he shook his head and whispered, "I really don't know where I am."

COVERING THE SACRED

The night before our Wainwright trip, Richard and I sat at the small wood table in the corner of his kitchen. Over a meal of venison burgers, he shared the story of Carlos Frank, a Koyukon Indian, who in 1975 was arrested and charged with killing a moose out of season. Frank argued the animal was needed for a religious ceremony. Two lower courts found him guilty. When the case got appealed to the Alaska Supreme Court, Richard was called in as an expert witness. The high court reversed the verdict, calling moose meat "the sacramental equivalent to the wine and wafer in Christianity."

The next morning, I woke early and eased down the creaky steps to make coffee. Richard came down when I was halfway through my second cup. He brewed cocoa and joined me in his quiet living room. We sat across from each other, enjoying the view. Low morning light shimmered through a veil of rain. A gang of crows squabbled over the mountain ash berries in the front yard.

"I was lying in bed this morning," Richard said, "thinking about that court case. I remembered Catherine Attla telling me that when you bring a plate of moose meat to your neighbor, you should cover it with a cloth to show respect to the animal's spirit. Catholics do the same thing. They cover the plate of crackers before communion.

"Every Sunday, a billion Catholics eat a cracker and sip wine in acknowledgment of their sustaining faith. Is it that big of a stretch to move beyond the metaphorical cracker—to accept that everything we eat and drink, whether it's peanut butter or moose meat, is a sacred exchange worthy of respect and gratitude? Are we really that far from seeing our way out of the mess we've made?"

In a flutter of wings, the crows flew out of the mountain ash. Richard watched them go and then gazed at the rain squall slipping across the bay. He took a sip of cocoa and went on.

"Catherine did not, in any way, see herself as separate from the world around her. She cared for the world as naturally as she cared for her own body. That's the perspective we need to regain," he said. "Our culture has moved so far from a sane relationship with the world. I believe the Koyukon worldview offers a way back."

After breakfast, Richard ran into town to check his mail. Alone in the house, I scanned the spines of books on the living room shelf—natural history and Arctic exploration. Not a single volume of fiction. Nothing he'd written was on display.

The shelves themselves, balanced on chunks of split firewood, were built from wave-worn planks. Stones and bones from far-flung beaches and rivers cluttered the top board. My gaze settled on the shattered and healed deer leg from Kruzof Island. I picked up the bone and ran my thumb over the shard protruding from the egg-sized lump.

I recalled Richard's words from the introduction to *Shadow of the Hunter*: "This book is not only about the North Slope people, it also *for* them. . . . It is always difficult to know where the trail is leading, but the difficulty is eased somewhat by knowing where it began."

I realized that as each generation inherits an increasingly depleted and worn-out world, the path forward, for all of us, becomes more difficult. Where do we turn to find the way forward? An honest look at my culture's tenure on this continent must confront the truth that the violence initiated by Columbus and his fellow sailors is not isolated in the past.

"We see a continuance in the present," wrote Barry Lopez, "of this brutal, avaricious behavior, a profound abuse of the place during the course of centuries of demand for material wealth. . . . Looking back on the Spanish incursion, we can take the measure of the horror and assert that we will not be bound by it. We can say, yes, this happened, and we are ashamed. We repudiate the greed. We recognize and condemn the evil. And we see how the harm has been perpetuated. But five hundred years later, we intend to mean something else in the world."

My friend Peter Forbes, after years striving to facilitate change, had concluded that a true shift "will only come out of the pull of love, joy, and restoration of healthy human life, rather than the push of fear."

I believed he was right. Yet fear is close and strong, stirred and stoked by each cycle of the news, each fresh report on shrinking ice and disappearing species. There is a real and justified call to arms.

It had been decades since Don Muller knelt down and locked his neck to a cleat, becoming one of Richard's heroes. Richard was likewise grateful for all protesters placing themselves before a pipeline, their voices rising up to say: *Enough! No more! We are right now, at this place, done with the greed. It is time to find a new way. Enough of our drills and saws and explosives tearing the tops off mountains. What the land needs now is our attention. We must cast off the lie that money makes us rich.*

I fingered that broken bone and looked out across the water. Shafts of sun stabbed a gun-gray sea. Ripples scurried across the bay, pushed by an invisible breeze.

What are the forces that heal?

How do we resist *and* keep telling the stories of love and joy and restoration? How do we move from harvesting resources to engaging in a sacred exchange? Where do we turn to replace wealth with wisdom, greed with gratitude?

BROTHERS

The Olgoonik Hotel was built to house the slug of workers prospecting for oil in the Chukchi Sea. Each cheaply paneled room was scarcely bigger than its one overly soft bed. Showers worked if you didn't mind cold water. We dropped our bags and went down to the cafeteria for breakfast. Dave, our driver, was seated with four coworkers. This late in the season, the remaining tables stood empty. The buffet line offered watery eggs, overcooked bacon, and warmed-over pancakes.

Halfway through our meal, a man slipped into the room to warm himself at a radiator just inside the hotel door. He was handsome, square-jawed, midfifties. Richard put down his fork and stepped across the room, hand extended.

"I bet I know some of your family," Richard said. "What's your name?"

"Jimbo. Jimbo Aveoganna."

"Well, sure, I hunted with your dad."

Jimbo joined us at our table. He was seven years old in 1964 but had fond memories of Niglik's time in the village. Over a cup of coffee, Jimbo shared the town's news—who had died, who was still living.

After breakfast, we shrugged into our coats and stepped into brilliant midmorning sun. The thermometer read –7 degrees. A steady breeze whistled off the water. Drifting snow snaked across the icy road. A snow machine zipped by, followed by a four-wheeler, driven by a kid with no hat, no gloves.

"Well? Where to?" I asked.

"Let's head east," Richard suggested. "Jimbo said the old part of town is that way."

Richard and Hank in the doorway of Richard's Wainwright home (Photo by Hank Lentfer)

We trudged past blocks of identical houses, a cluster of snow machines pulled tight to each one. Within a quarter mile, the regiment of newly constructed homes gave way to the haphazard placement of older dwellings. A lone husky, chained to a post, howled as we passed.

"Magnify that sound five hundred times," Richard said. "That's what this place used to sound like. Hard to believe that's the first dog we've seen."

As the road arced east, homes got smaller, the yards bigger.

"I'm starting to recognize things now," Richard said. "There's the old Assembly of God church, which means my house should be right— Hey, look! There it is."

A hundred feet from the road stood a structure wrapped by drifted snow. Our boots squeaked on the wind-packed crust. The door hung from one hinge. Flakes of dull yellow paint clung to weathered gray siding. Richard walked past the house to the nearby bluff. Gray clouds blended into dark-gray water. Small waves pushed a line of ice up the black sand beach.

"It's weird. Tagruk's house used to be right back there," Richard said, gesturing away from the eroded bank. "My house was well back from the shore. Now it's beachfront." He spun a slow circle, trying to make sense of the changes. He then moved toward the skeleton of an upturned boat resting on a wooden rack. Richard ran his hand over the cracked ribs.

"This must be Tagruk's umiak," Richard said. "We hunted walrus and ugruk from this boat. In the spring, we put it on a sled and pulled it to open water to hunt whales." Richard walked around the boat. He fingered tattered remnants of hide lashed to the splintered gunwale, then gazed across the wind-flecked waves.

He tramped back to the house and pushed aside the sagging door. A finger of blown snow stretched into the room's dim interior. Cans and bottles lay strewn around an overturned table. A refrigerator, its door torn off, stood along one wall.

Richard seemed blind to the mess, animated by memories.

"This is the same stove," he said, touching a black knob on the dingy appliance. "I cooked vats of dog soup on this thing. And look up there. I hung a kayak from those hooks. Tagruk helped bend the ribs. Ikaaq helped sew on the skins. You wouldn't believe the stink of those hides.

"Oh my God. Listen to that!" Richard pointed to a moisture vent high on one wall as a gust of wind rattled the small plywood cover. "I can't believe the *qingaq* is still making that noise. That was my weather report. I'd listen to that rattle to gauge the force of the wind. When the storms really got howling, this whole place would rock and shake. I'd lie in bed, convinced this house was about to tumble across the tundra. In the deep cold, the earth cracked. Explosions in the middle of the night, like gunfire. I could never get back to sleep when that happened."

After an hour of reminiscing, we stepped back into the bright sun. A snow machine pulled into the neighboring yard. Richard approached as a young woman swung her leg off the sleek machine.

"Excuse me," Richard said. "Can you tell me where Tagruk lives?"

"Right up there. Follow this road. Yellow house with the big window."

"Didn't he used to live right there?" Richard asked, pointing to a spot nearby.

"Yeah. But they have to move all the old houses. The sea keeps coming closer and closer."

In planning the Wainwright trip, I suggested we invite a videographer. "A researcher going back after fifty years—that's a rare thing," I said. "There's so much to learn."

Richard never even considered it.

"I'm not going to make this a media event," he said. "I'm just going to see some old friends."

As we neared Tagruk's house, I reached into my pocket and flipped on a small digital recorder. I wanted to at least capture audio of the reunion.

"You're going to have to rely on your memory," Richard said. "We're not taking anything—no photos, no recordings. I'm just here to visit."

There was a yellow two-door SUV parked alongside a couple of snow machines in Tagruk's driveway. No dogs. To the left of the front door stood an upturned oil drum. The body of a single goose lay atop the drum. Beneath a thin dusting of snow, I made out the distinctive white-on-black necklace of a brant goose.

"Look at that," I said. "A niglik."

"Sure enough," Richard said.

The outer door was open. We stepped into a narrow entryway. A tangle of steel traps hung alongside a bundle of parkas and snowsuits. The hindquarters of a caribou lay on bloodstained cardboard—dark meat laced with pure-white fat—a heart and tongue alongside the haunch.

Richard knocked on the inner door. A woman answered, two hands on a walker, round cheeks pushed wide with a full-toothed smile.

"Mabel!" said Richard.

She set her walker to the side and gave Niglik a hug. Settled with age, her head just reached the bottom of Richard's ribs. Tagruk appeared, gripped Richard by the shoulders.

"Azaa, look at you," Tagruk said. "You got old."

They held each other's shoulders and locked eyes, laughing and smiling at the deep pleasure of being together once again.

"You look just like your dad," Niglik said. "Laugh like him too."

Tagruk's wife shuffled to the kitchen table. Tagruk put the kettle on the stove. He was grizzled and gray but moved about the tiny kitchen with a straight back and sure feet. He poured black tea into white enamel cups.

After a while, Richard brought out the stack of black-and-white photos. The top image showed six men, each seated on a walrus, each sharpening a machete or an ax, the ice at their feet stained with blood. An umiak floated alongside the ice floe.

"That was a good hunt," Tagruk said. "Best luck ever. Got nineteen that time. That was enough. We gave you one to cut up all by yourself. We wanted you to learn."

"Look at this one," Niglik said. "This is the kayak we built. I remember you helping me bend the baleen ribs right there in my little house."

"Yeah, yeah. We had a lot fun. You didn't know how to do a lot of things, but you knew how to laugh."

"And this one. It's Kusiq. Your dad's looking for a polar bear from the top of *piqaluyiq*."

"You don't see ice like that anymore."

"And here's a picture of my dog team."

"I remember," Tagruk said, "your dogs were really slow, but they were like you—never give up."

The stories flowed well past the end of the photos and the last of the tea. Toward evening, we stood and reached for our coats, promising to return the following day. At the door, Tagruk again gripped Niglik's shoulders. "I love you," he said. "You are like my brother."

Richard nodded, his voice too gripped with emotion to say anything in return.

PATHS OF PRAYER

Florence Nelson wanted her boy Richard to be a Lutheran. When she felt her adolescent son straying from the flock, she invited the pastor for tea and cookies. Had that pastor, just one time, used his position at the pulpit to extol the beauty of turtles or rave about the secret lives of salamanders, he might have had a chance. Instead, out of respect for his mother, young Richard nibbled his cookie and listened to the pastor ramble on about eternal rewards. He then went out to hunt for snakes.

The curiosity that drew Richard to prowl Madison's swamps eventually lured him to the Chukchi Sea. After running dogs along the Arctic coast, he mushed over the frozen ribbons of the Black River and the Kobuk. On the banks of the Koyukuk River, a new dimension opened when Richard witnessed the Attlas moving through a forest of eyes. The trees, the river, moose, redpolls, geese, even the wind, watched and measured their steps. For Catherine and Steven, being embraced by the web of life was, like the presence of oxygen, an unquestioned reality. They were immune to loneliness.

Richard envied his Koyukon teachers "for the surety and comfort of their knowledge, and for the gift of intimacy with nature that my own ancestors let slip away. I'm grateful for what I've learned, but sad over what I've lost, and troubled by my abiding doubts."

There were no worn paths of prayer that led Richard to the shores of Kruzof Island. He boated across uncharted waters, drawn by the same yearning that compels others to light a candle or reach for a rosary. Richard went to the island seeking a spiritual path and escaping heartbreak. Hunting, surfing, and watching became his meditation. Writing, his discipline. The rain forest, his cathedral. The deer, his communion. After days, weeks, and months "rooted in the unshakable sanity of the senses," Richard was struck by the absurdity of our culture's insistence in dividing the world into separate bits: self/other, male/female, wild/tame, mine/yours, sacred/profane, organic/inorganic, plant/animal, human-made/natural. In the Rain Journal, he expounded upon these false divisions:

> *We fail to recognize such categories are an artifice and the customs surrounding them are pure invention. Throughout human history the artifice of boundary has been a locus for conflict, death and enslavement, ranging from the conflicts among individuals and families to world wars. And in the recent history of our own culture we have built a new border between humanity and the rest of creation. This may be the most dangerous and threatening boundary of all, and the most artificial.*
>
> *We have only to think how ephemeral these boundaries and categories are to realize the folly of investing too much in them. A salmon rots away at its own boundaries even before it dies and dissolves completely into the stream. Where are the allies and enemies when they have moldered back into the soil? Man and woman, black and white, animal and plant, human and non-human, wild and tame—all that is quick and hot will cool away, washing into the earth, the boundaries gone.*

In the thirty years since I first met Richard on the shores of a forested lake, I've come to see my initial impulse was right.

He is a monk.

There are no robes. No frock. No cross. Just jeans, T-shirt, and a ball cap. He hates doing his taxes, rages at his computer, swears like a sailor, and eats a peanut-butter-and-jelly sandwich almost every day. Yet he's dedicated his life to seeing the invisible forces binding us all together. In a culture lost in the delusion of separateness, he's become an unlikely, irreverent holy man. His awareness makes him grateful, which makes him laugh, which makes others want to share his joy.

Richard went to the woods seeking solace for a bruised and cautious heart. He returned, years later, knowing "the rightness of what Koyukon elders teach—that no one is ever alone, unseen, or unheard, and that gratitude kindles the very heat of life."

Who are these people?

What is this land?

Richard holds these questions as one would a flashlight on a starless night. Stories from a distant time, a wailing loon on a black lake, the raining clicks of a thousand caribou hooves—every place, person, and creature that falls under the light of these probing questions gives him a fresh answer.

Richard might have pursued tenure or planned for retirement had he not been so focused on a more enduring source of wealth. Instead, he devoted his days to wild country and the people who know it best. He mushed and paddled and hiked and floated and spent more nights sleeping on the ground than anybody I know. Along the way, he discovered the truth of his friend Barry Lopez's words: "True wealth—sanctity, companionship, wisdom, joy, serenity, these things were not to be had without an offer of heart and soul and time. . . . The true wealth that America offered, wealth that could turn exploitation into residency, greed into harmony, was to come from one thing—the cultivation and achievement of local knowledge."

Through years of practice, Richard became a patient, voracious listener, bearing joyous witness to the world's beauties—light glinting through a dragonfly's wing, jellyfish pulsing beneath a rolling sea, the final sunset of an Arctic winter.

And he did it all with an abundant offering of heart, soul, and time.

"'Remember you are not alone,'" said the Dalai Lama in *The Book of Joy*, "'and you do not need to finish the work. . . . It helps no one if you sacrifice your joy because others are suffering. We people who care must be attractive, must be filled with joy, so that others recognize that caring, that helping and being generous are not a burden, they are a joy. Give the world your love, your service, your healing, but you can also give it your joy. This, too, is a great gift.'"

A sure sign that Richard did, indeed, discover North America's true wealth has been his irrepressible desire to give it all away. Through words on the page and the stage, through recordings and films, he's dedicated every fiber of his artistic impulse to giving voice to the lingering beauties and subtle wisdom to be found in a tattered land.

Who are these people?

What is this land?

The specific answers are less important than the act of inquiry. The questions elicit a posture of discovery and convey a notion of respect. They form an efficient tool for uncovering the riches of any given day. We learn of our place in the world as soon as we quit thinking about ourselves.

Paul Ongtooguk is an Iñupiaq from Nome and a professor of indigenous education at the University of Alaska Anchorage. He has described much of the writing about Inuit culture by outsiders as "accurate, but false." When a student asked for his thoughts on Richard Nelson's work, Ongtooguk said, "Ah, Richard Nelson . . . My father and his brothers were talking about him one day, trying to figure him out. They decided that somehow an Iñupiaq wandered far from home and died in the Midwest. His soul was reborn in Richard Nelson, who finally made his way back home."

FINAL GIFTS

News of Richard's return flashed through town. People stopped by the hotel and pulled over on the streets. In between the impromptu visits, Richard sought out old friends, knocking on the door of every old-timer in town. Although none of these men still hunted, each home had its own bloodstained cardboard and chunk of fat-rich caribou. The meat was gifted from grandchildren, nephews, and neighbors. One of the young men supplying his elders with meat was a grandson of Kusiq Bodfish—named Niglik.

Niglik stopped by the hotel to meet the white man who shared his name. In 1964, every conversation had eventually circled around to hunting. And, now, sitting in the Olgoonik cafeteria, the talk, once again, turned to caribou.

"Got three yesterday," young Niglik said. "I was checking my nets at the Kuk River when they came by."

"I hear there are salmon in the river now," Richard said. "Never used to be."

"Yeah. I'm catching more and more pinks. Even some cohos. Once in a great while, I get a king."

"It's not easy running nets under the ice."

"I'm the only one now. My granddad taught me how before he died."

That afternoon, Chester Ekak, grandson of Kusiq's friend Ikaaq, stopped in for a visit. He wanted to meet the anthropologist who was a part of so many of his grandfather's stories. In his midthirties, Chester bristled with compact, intense strength. He cast a keen eye on Richard's photos and asked sharp questions about each man and every boat. He then pulled a phone from his breast pocket and shared photos of his own—sledloads of caribou and bundles of ducks. He flipped past pictures of his children and settled on an image of a bowhead whale rising alongside an ice floe.

"Let me tell you a story about hunting whales with your grandfather," Richard said. "It was springtime. We were camped on the ice beside a small patch of open water. Over a week, we'd seen only two whales, both too far away to bother launching the umiak. I remember we were all sleeping in the wall tent when your grandfather said, 'Whale coming. It will be close to the ice.' Everyone rushed out and stood alongside the umiak, ready to launch. A few minutes later, a bowhead's breath erupted one hundred feet from the ice.

"I couldn't fathom how your grandfather knew that whale was coming. He was in the tent with the rest of us—no way he could have heard or seen it. When I asked the other men how he did it, they just shrugged and said, 'He just knows.'"

Chester nodded. Richard went on.

"When I came here, your grandfather was already an old man. But he was always watching. At whale camp, he spent days observing the ice move. When a whale dove, he'd keep track of how long it was down and how far it had traveled before resurfacing. He was the greatest hunter in town, and yet he was still studying, still learning."

"I'm a whaling captain now."

The simple statement was Chester's way of acknowledging he understood the intertwined roles of leadership and learning. Chester then described the skills and character of his crew. He spoke with affection and pride, bragging about the strength of his harpooner, telling of near misses, sudden storms, thick fog, and the careful choreography of successful hunts.

His stories stirred my own childhood memories. I felt, once again, the weight of a whale, the pride of pulling, the satisfaction of a community in need of my strength.

These days, by spring whaling season, the sea ice is thin, making camping on the edge a treacherous affair. Last year, Chester's crew struck a bowhead in the nearly ice-free days of fall. They had to travel forty miles offshore to find a whale. It took the combined pulling power of six boats two full days to tow the bowhead to shore. "It was a race against time," Chester said. "We knew that whale was rotting from the inside out."

Richard recalled his prediction that Iñupiaq knowledge would be "lost forever in the icy graves of the old men." He had, for five decades, assumed he was right.

It is true that Chester navigates a world his grandfather would not recognize. The intricate knowledge of hunting seals along leads and breathing holes is melting with the sea ice. With snow machines outfitted with GPS units, there is no need to navigate by the stars. Some knowledge has indeed been lost. Yet in a world of F-350 pickups and big-screen televisions, Chester still defines himself as a hunter. Young Niglik too. They see searching for animals as the weaving of connection, to themselves and their ancestors, to food and place, the past and the future. The ice is shrinking, but the urge to pull together remains.

"I've got something for you," Richard said. "Give me a minute."

He returned from his room with a copy of *Hunters of the Northern Ice* and *Shadow of the Hunter*. Chester accepted the books, held one in each hand.

"You really wrote *two* books about Eskimos?"

"No. I wrote two books about the people of Wainwright. Your people."

Chester opened *Hunters of the Northern Ice* and read the inscription:

For Chester,

What a great pleasure to meet you and learn that I was wrong, to see that hunting did not get buried in the icy graves of the old men.

Niglik

When Richard arrived in 1964, he'd only dreamed of living in a community where every person was a student of the earth and sea, weather and animals, moods and migrations. The hunters' teasing might not have cut so deep if he hadn't so desperately wanted to belong. Fifty years later, seated on a bed upstairs in the Olgoonik Hotel, he thought it all through.

"This place, these people, mean so much to me," Richard said. "I never dared believe my presence here somehow mattered to them. I always thought they viewed me as just some dumb guy who is fun to be with but will kill himself if we don't teach him something." He lay down, clasped his hands behind his head, and gazed at the ceiling. "I always felt people here

Richard sharing a laugh with Tagruk and Mabel Bodfish in their Wainwright home (Photo by Hank Lentfer)

were generous. But this time, they're assertively affectionate. These people gave me so much. I didn't know how gratifying it would be to be able to give a little something back."

We reserved much of our last day for a final visit with Tagruk and Mabel. Tagruk filled the same enamel cups with more bitter tea. We took our places around the table. Two things I will never forget happened that afternoon: One I'd anticipated. The other was a great surprise.

During our first visit, Richard had shared stories about my life as a means of introduction. In the mid-1960s, the same years Richard lived in Wainwright, my family lived ninety miles east, in Utqiagvik. As a polar-bear biologist, my father had traveled along the coast and spent time on the ice with Tagruk's father, Kusiq. At the time, I was just a baby. But seated in Tagruk's quiet kitchen, I realized it mattered that my early memories were formed on the sea ice.

That first day, during a lull in the stories, Tagruk turned to me.

"You got Eskimo name?"

"No, I don't."

The subject didn't come up again.

On that final day, the kitchen clock ticked in the space between the end of one story and the beginning of the next. Tagruk drummed his fingers on the table, then turned my way.

"Kusiq. That's your name now."

I'd done nothing to earn the honor. For five days, all I'd done was quietly sip tea and listen. Calling me Kusiq was a gift to Richard, a way to say, "You're always welcome here. Your friends are my family."

Everyone around that table knew these men would never see each other again, that whatever was said in that quiet kitchen would be their final exchange. This was Tagruk's last chance to give something to his friend and brother, the man who'd learned to mush dogs and skin caribou, to hunt seals and walk on thin ice, the man who, after all these years, had come back.

The weight of the gift welled up in Richard's eyes. Mine too. Even Mabel took off her glasses and caught a tear with the corner of a napkin.

And in that gentle space, Richard asked for something more. "I was hoping you'd sing that little niglik song. I want to hear it one more time. Would you sing for me?"

Without as much as a breath, Tagruk began. He sang in whispered tones, soft words in a lilting melody.

> *Nigligaichguk tunnguraalarut, tunngiich aasii kangiqsilyaangmigaich.*
>
> The brant goose speaks in English, but the white man doesn't understand.

The words, of course, were indecipherable to me. But the burst of laughter at the song's end, I understood. It was the laughter of love and friendship. But it was more than that too. What had started as tender teasing all those years ago had grown into a shared bit of music. Both men were now in on the joke. They laughed together, two old hunters, chuckling with the knowledge that anyone can come to understand.

You just have to learn to listen.

AFTERWORD: WINGS

Three days of bone-chilling rain and then, last night, the clouds thinned to reveal a veiled moon and faint stars. This morning I drink coffee in the pre-dawn quiet and prep my gear for another day in the woods. I've come to the island to hunt—at least that's what I'd told my family and friends. And, true enough, I do carry a rifle, but the search for deer is an excuse.

The forest is a refuge, the feathered branches and high twitter of crossbills a soothing change from the tangle of tubes and beeping machines of the intensive care room. Nels's doctors were wise and kind, his nurses skilled and caring. When they said there was nothing left to do, I joined a band of friends gathered around his bed. The machines were silenced and the lights dimmed. Someone brought candles.

Birdsong played from a speaker perched on the windowsill. Nels, years before, had created the recording, crafting an intricate soundscape of his rainforest home. We humans went quiet. We let the warblers, sparrows, and thrushes speak.

I remember robin song, free and clear, interrupted by a raven's raucous call. The raven called again, sharp and insistent, as Nels drew a final shuddering breath.

There were plenty of tears in those final days but, for me, at the end, sadness was displaced by (dare I say it) an expansive joy. And I've found few tears here in the forest. I move slowly beneath the towering spruce and hemlock. I pause to watch the tense twitch of a Pacific wren searching a downed log for insects, its soft, chipping call loud in the still woods. I smile at the friendly banter of chickadees twittering in the branches of a young hemlock. And ravens. Each day they find me. They swoop beneath the canopy, air huffing through wings. They cock their head, peer down, land on a branch, and prattle and croak the day's news.

I know, in the coming days, months, and years, I'll miss my friend. I might pass one of our favorite campsites and flash back to a breakfast of cereal and laughter, both of us drunk on the morning's beauty. Or I'll reach for the phone, eager to share the news of wolves in the yard or owls in a tree. But right now, he's too close to miss. I feel him in the damp moss, sense him moving with the elusive deer, imagine him twisting out over a gray sea testing his shiny black wings.

Yesterday a buck stood close and still. I hauled his body back to the cabin, leaving the entrails on the moss, a feast for the ravens. I'm grateful for the deer's life, the flesh of his body will nourish my own in the months ahead. And I'm grateful for Nels's life. His tender love and deep laughter have enriched my days. His lessons on listening will nourish my soul in the years to come.

I've come to the island by myself but I am not alone.

Such is the gift of raven's world.

Hank Lentfer
Tàas' Daa
November 2019

ACKNOWLEDGMENTS

It's a strange thing making art out of someone else's life. Especially when that someone is a dear friend. Nels allowed me to paw through precious photos, interview old flames, and read thousands of pages of his personal journals. I thank him for his trust. And I thank him for the warmth of laughter on cold mornings. Nels has taught me how to be more fully alive. There is no way to give adequate thanks for such a gift.

Throughout the years of research and writing, I've settled into an uneasy place in an unlikely lineage of storytellers. I am grateful for so many people I never had the chance to meet in person: Catherine and Steven Attla, Lavine Williams, Kusiq Bodfish, Wesley Ikaaq, among so many others. Unable to seek consent from people no longer living, I've risked telling stories that are not mine to tell. Trying to be respectful does not mean I have succeeded. I cannot undo my mistakes; I can only pledge to do my best not to repeat them.

In the last years of his life, Nels adventured, traveled, and shared his home and heart with Debbie Miller. Through the highs and lows of cancer, Debbie was right there—a devoted advocate, a tireless researcher, a perpetual optimist, and a loving partner. I am grateful my friend was blessed by Debbie's comfort and care.

Every paragraph of *Raven's Witness* is shaped by what was left out. Nels's life floated on the love of friends with whom he surfed, hunted, explored, and shared Thanksgiving (the man's favorite holiday). Neither the omission nor inclusion of people in these pages reflects an individual's importance in Nels's life. To adequately convey the potency of his core friendships would have swelled this book to five times its current size.

In person or over the phone, the following folks generously agreed to be interviewed: Angela Gonzalez, Annie Caufield, Barry Lopez, Bill Schneider, Chip Blake, Dave Nelson, David Brent, Don Muller, Gary Nabhan, Gary Snyder, George Gmelch, Jerry Burgette, John Straley, Jonathon White, Kathy Mautner, Ken Fate, Ken Taylor, Marion Gilliam, Mark Badger, Nita Couchman, Ray Bane, Rick Bass, Rick Caulfield, Sam Skaggs, Scott Russell Sanders, and Sharon Gmelch. I thank each person for their time and insights.

Deep gratitude to Barry Lopez for the time and thought devoted to the book's foreword.

Special thanks to Robert Osborne, ardent fan of all things Alaska and wild, for his generous support of our trip to Wainwright.

Deep thanks to Lauren Oakes, the smarter half of our two-person writers group, for edits to the first clumsy draft of each chapter. Thanks to Kim Heacox, close neighbor and dear friend, for his disdain for sloppy writing and his faith in the power of a properly chilled and well-timed beer. Thanks to my friend Peter Forbes for the never-ending conversation about the restorative power of story. Thanks to Helen Whybrow for a thorough edit of the first draft. The following folks read parts or the entirety of various drafts: Debbie Miller, George Gmelch, Laura Marcus, Richard Nelson, and Zach Brown. Thank you all for your gifts of time and attention.

Kate Rogers, editor in chief at Mountaineers Books, kept in touch through the months of writing. Kate's insight and judgment through the final shaping of the book have been invaluable. And deep thanks to the dedication and artistry of Helen Cherullo, Mary Metz, Jen Grable, and the entire staff of Mountaineers Books. It's been a comfort to have this manuscript in the hands of such a competent, caring team.

My daughter, Linnea Rain, wrote a novel while I worked on this book. Mornings began with me at my desk and Linnea perched on the nearby bed. I thank Linnea for her companionship through all those mornings quietly chasing words.

This project, like everything I do, has been made possible by the ever-renewing benefits of the Anya Maier Fellowship of Love. I know of no words capable of expressing the depth of my gratitude for Anya's unwavering support, unconditional love, and shared affection for long days in the quiet and rain-drenched forests of home.

APPENDICES

I. PARTICIPATORY ANTHROPOLOGY

Richard's doctoral work focused on the lifeways of the Gwich'in Athabaskans of Alaska's boreal forest. He'd spent a year living in Chalkyitsik (population ninety-five), an isolated village on the edge of the Yukon flats. Richard asked few questions of his neighbors there. During his time in Wainwright, he'd learned that participation was a far sharper tool than inquiry. Within weeks of his arrival in Chalkyitsik, he had dogs staked in the yard, fresh wood stacked against the cabin.

Richard left for Santa Barbara in June 1970, where he rented a small house in the suburbs and hunkered down to write. He worked, nonstop, for 210 days. With a sore back and frayed nerves, he emerged from his cottage and delivered one thousand typed pages to his doctoral committee.

A week after Richard handed in his dissertation, Dr. Tom Harding summoned him. The professor's office was cluttered and cramped. Packed bookshelves spanned floor to ceiling. A single window opened to a second-story view of students strolling across campus. Harding leaned back and crossed his legs. He pressed his fingertips together and gazed at the thick typed manuscript centered on his desk. Richard sat on a metal chair across from his professor. He twirled a pencil in his fingers and waited.

Dr. Harding uncrossed his legs, leaned forward, and tapped the manuscript.

"This is more than anyone ever wanted to know about hunting moose."

Richard flushed red but held his tongue.

Dr. Harding continued. "This manuscript needs to be cut in half. If you want a degree from this university, you've got to move beyond pure description and incorporate some theory."

Richard shot to his feet. "I don't need this," he snapped. "I don't need any of this. I'd be happy to be a farmer." He bolted from the office, scuttled down the stairs, and strode across campus.

Hunters of the Northern Forest was, in Harding's eyes, a scholarly ethnography. For Richard, the manuscript was much more than an academic work; heavy with the isolation and effort of village life, it encompassed trees felled for the hungry stove, drinking water hauled from the river, rabbits snared to feed the dogs, moose meat hauled for the neighbors, traps checked at fifty below, sleepless nights devoted to field notes.

The heat of Richard's anger reached beyond Santa Barbara. As a master's student in Madison, Richard had delivered a detailed lecture about Iñupiaq hunting techniques. Afterward, the professor addressed the class: "This is all great information, but who here thinks this is actual anthropology?" The question, following the stress and loneliness of Wainwright, cut deep. If risking his life and enduring ridicule to document details of Arctic living was not enough to satisfy the lords of anthropology, what more must he do?

But Dr. Harding was right; the dissertation was too long.

Richard was wrong; he didn't really want to be a farmer.

So he returned to his desk. He cut the manuscript in half and reluctantly added chapters comparing the Chalkyitsik Kutchin to Iñupiaq Eskimos.

The required inclusion of anthropological theory strengthened Richard's conviction about the importance of detailed description. In the introduction to *Hunters of the Northern Forest*, Richard notes that just as one cannot learn to ride a bike through reading, there are subtle and invisible complexities to subsistence activities that only come into focus through full-on participation.

> *It is possible to learn a great deal about setting traps by listening to descriptions of the techniques given by expert trappers; but somehow the accounts never tell as much as being right there to watch*

> *trap sets put together. And even after watching the same process time and again, you are almost certain to make a mistake when a man hands you a trap and says, "Set it over there." How long were the twigs he used here? Does the trap belong this far inside the cubby? How high did he place the bait? Did he push the spring against his knee when he set the trap? You put the set together, your instructor comes over, looks, and starts moving things around. One or two more tries and you have it right. Now you are ready to write a description of a trap set. One never realizes how little he knows until someone says, "Now you try it."*

When *Hunters of the Northern Forest* was reissued thirteen years after initial publication, Richard took the opportunity to write a retrospective.

> *I have . . . chosen to delete the two final chapters from this book. . . . The fundamental purpose of this study has always been to present a descriptive account of Kutchin subsistence life. . . . Comparative judgements and theoretical speculations do not contribute to its central purpose. . . .*
>
> *Over the years since* Hunters of the Northern Forest *was written, my personal perspectives on the goals of ethnographic study have evolved, or at least have settled comfortably on a position I once lacked the confidence to defend. I have become convinced that purely descriptive ethnography is valid and essential in its own right. . . .*
>
> *In fact, there may be few achievements more difficult than to portray perceptively and evocatively the richness of another cultural experience. . . .*
>
> *We should recognize that our ability to translate the human experience into words and images is still rudimentary. The mass of detail encompassed within a cultural moment is literally beyond reckoning; and the degree of subtlety in each fragment of human behavior strains the limits of our senses. . . .*
>
> *Anthropologists are aware that good description is virtually timeless, while preferences in theory are as changeable as the winds that blow through Kutchin country.*

Decades later, Richard remained a staunch proponent of experiential learning and a sharp critic of some anthropologist's insistence on "*explaining* why Native people organize their cultures as they do—as if it is the white man's burden and privilege to understand all." Richard believed the highest role of anthropology "is to take the wisdom and lessons of other cultures back to their own people, to enlighten them and change their lives for the better, to make the exchange of cultural richness a two-way process."

II. BOOKS AND MONOGRAPHS

This is a complete list of Richard's published books and monographs.

Books:

Hunters of the Northern Ice. Chicago: University of Chicago Press, 1969.

Hunters of the Northern Forest: Designs for Survival Among the Alaskan Kutchin. Chicago: University of Chicago Press, 1973.

Shadow of the Hunter: Stories of Eskimo Life. Chicago: University of Chicago Press, 1980.

Haunted by the lack of humanity in *Hunters of the Northern Ice*, Richard sat down to write a piece of fiction. It was a risky move. He had no publisher. Scientists were not supposed to make things up. But he felt compelled to write a descriptive account that "emphasizes the raw material of the senses." *Shadow of the Hunter* is an inspiring example of the scholarly brought to life through story.

The Athabaskans: People of the Boreal Forest. Fairbanks: University of Alaska Museum, 1983.

Make Prayers to the Raven: A Koyukon View of the Northern Forest. Chicago: University of Chicago Press, 1983.

Richard wrote his first two ethnographies in the third person, the clinical point of view of social scientists. *Make Prayers to the Raven* mixes this standard approach with first-person journal entries. Richard's commitment to participatory anthropology made the study of Koyukon spirituality a deeply personal affair. No longer could he write about a changing culture without reflecting on how the culture was changing him.

Interior Alaska: A Journey Through Time. Editor and coauthor. Seattle: Alaska Northwest Books, 1986.
The Island Within. Berkeley, CA: North Point Press, 1989.

In 1964, Richard helped the US Air Force write a survival manual for pilots stranded on sea ice. *The Island Within* is a survival manual for a society estranged from the natural world.

Heart and Blood: Living with Deer in America. New York: Alfred A. Knopf, 1997.

After years living with Alaska Natives, Richard focused his anthropology skills on the backwoods of his own culture. He followed Wisconsin hunters on opening day, rode along with Texas ranchers growing trophy bucks, and lived with researchers trying to unravel the social lives of deer. The structure of *Heart and Blood*, a mix of journalism and personal reflection, reflects Richard's journey from ethnographer to poetic writer.

Patriotism and the American Land. Coauthored with Terry Tempest Williams and Barry Lopez. Great Barrington, MA: Orion Society, 2002.

Richard's essay in this three-author anthology is drawn from the talk he gave at the 1999 Fire & Grit gathering. Publishing in partnership with Terry and Barry, he told me, was the peak of his writing life: "I am so grateful for the community of writers. The essence of what I have to say is poured into that piece."

Published Monographs:
Alaskan Eskimo Exploitation of the Sea Ice Environment. Fort Wainwright, AK: Arctic Aeromedical Laboratory, 1967.
Literature Review of Eskimo Knowledge of the Sea Ice Environment. Fort Wainwright, AK: Arctic Aeromedical Laboratory, 1967.

Kuuvangmiut Subsistence: Traditional Eskimo Life in the Latter Twentieth Century. Coauthored with Douglas B. Anderson et al. Washington, DC: National Park Service, 1977.

Tracks in the Wildland: A Portrayal of Koyukon and Nunamiut Subsistence. Coauthored with Kathleen H. Mautner and G. Ray Bane. Washington, DC: National Park Service, 1978.

Harvest of the Sea: Coastal Subsistence in Modern Wainwright. Barrow, AK: North Slope Borough, 1982.

III. JOURNALS

Wainwright Journal

The Wainwright Journal covers two trips to the village: September 1964 through May 1965, and May 1966 through August 1966. An aspiring social scientist, Richard journaled with an objective tone, focusing on sea ice, animal harvests, and natural history.

Chalkyitsik Journal

Richard lived in the Kutchin village of Chalkyitsik from August 1969 to July 1970, doing field work for his doctoral dissertation. Each day's entry typically begins with a few paragraphs about the day's weather and logistics. Richard then broke subjects into discrete categories, with labels like BEAVER, TRAPPING OF or MOOSE, TANNING HIDES. While writing his second book, *Hunters of the Northern Forest*, he took a pair of scissors to a copy of the journal and pasted entries onto five-by-seven index cards. Organizing cards by subject created the book's structure.

Kobuk Journal

Richard lived with Kobuk Eskimos from December 1974 through April 1975, while under contract with the National Park Service. In addition to ethnographic information, this journal includes long philosophical explorations about the effects of changing land ownership on traditional lives.

Huslia Journal

Richard lived with Koyukon Indians from October 1975 through June 1976, again under contract with the National Park Service. In addition to weather and tightly organized ethnographic information typical of earlier journals, Richard introduced a diary section (labeled as such) into each day's entry. The diary explores his emotional response to village life and his changing awareness of the spiritual dimensions of the natural world. As an advocate of participatory anthropology, studying the spiritual side of Koyukon life, Richard tried to understand what he was learning at a personal level.

Rain Journal

Richard began the Rain Journal on January 1, 1984. For each of the next 5,110 days Richard, without fail, made time for the journal. Beyond a weather summary beginning each entry, this journal shows little stylistic allegiance to previous journals. The anthropologist trying to make sense of another culture is replaced by a man striving to understand his relationship to family and place. The Rain Journal is honest, vulnerable, and deeply personal.

IV. TELEVISION, RADIO, AND FILM

Television:

Richard, in partnership with producer Mark Badger, wrote *Make Prayers to the Raven*, a four-part documentary television series about Koyukon Indian relationships to the natural world. The show was broadcast in the United States on PBS in 1987, in Canada on CBC, and worldwide on BBC.

The idea for this series began when KUAC, a Fairbanks-based public radio/television station, received a grant to create a documentary about Koyukon lifeways. The station wanted Richard to be the on-camera anthropologist. Richard, uneasy posing as an expert on someone else's way of life, had a different idea. He proposed placing control of the film in the hands of a Koyukon council of elders. In addition to deciding what would be filmed, the council would have complete editorial control. It was a novel, risky approach. Mark Badger, however, thought it was a great idea, and his enthusiasm convinced the initially skeptical station manager to give it a try.

Richard pitched the idea to Steven and Catherine Attla in Huslia, Lavine Williams in Hughes, and other friends and acquaintances along the Koyukuk River. He arranged for Wayne Attla, the son of Steven's brother, to work as Mark's soundman.

For two years, Richard and Mark lived in service to the documentary. When the council called, they'd travel to Huslia—Richard from Sitka, Mark from Fairbanks—and record a funeral service, a bear hunt, whatever they were told. Mark edited hours of footage. Richard wrote the final script, which was narrated by Barry Lopez. Neither Richard's image nor voice appears in the TV series.

"Helping people craft the stories they want told—this is how anthropology should be done," Richard told me. "It is not an outsider's role to decide how to portray a culture. I wish I'd been taught this technique early in my career."

Radio:

After giving up writing, Richard turned to radio. For ten years, he wrote, narrated, and produced *Encounters*, a weekly half-hour radio exploration of the natural environment. From albatrosses to wood frogs, with over one hundred topics in between, each episode has a sharp focus imbued with Richard's unbridled enthusiasm.

The shows were broadcast on public radio stations throughout Alaska, with additional broadcasts on National Public Radio and the Australian Broadcasting Corporation. The shows are available as podcasts at www.encountersnorth.org.

Film:

For the last twenty years of his life, Richard had a creative partnership with Sitka-based filmmaker and writer Liz McKenzie. They collaborated on short films and essays that celebrate, educate, and explore our relationship and responsibility to the natural world. Their work culminated in the production of *The Singing Planet*, a thirty-minute film exploring the power and beauty of wild voices. All of their work can be found at www.encountersnorth.org.

V. ARTICLES, ESSAYS, CHAPTERS, EXCERPTS, AND INTERVIEWS (PARTIAL LIST)

"Relationships Between Eskimo and Athapaskan Cultures in Alaska: An Anthropological Perspective." Supplement, *Arctic Anthropology* XI (1974).

"The Inuk as a Hunter." In *Inuit Land Use and Occupancy Project: A Report*, edited by Milton Freeman, vol. 2, 203–206. Ottawa: Department of Northern and Indian Affairs, 1976. (From *Hunters of the Northern Ice.*)

"Hunters of the Northern Ice." In *Custom Made: Introductory Readings in Cultural Anthropology*, edited by Charles C. Hughes. Chicago: Rand McNally, 1976. (From *Hunters of the Northern Ice.*)

"Cultural Values and the Land." In *A Study of Land Use Values Through Time*. Fairbanks: Cooperative Park Studies Unit, University of Alaska, 1978.

"Athapaskan Subsistence Adaptations in Alaska." In *Alaska Native Culture and History*, edited by Yoshinobu Kotani and William Workman. Osaka, Japan: National Museum of Ethnology, 1980.

"A Conservation Ethic and Environment: The Koyukon of Alaska." In *Resource Managers: North American and Australian Hunter-Gatherers*, edited by Nancy M. Williams and Eugene S. Hunn. Washington, DC: American Association for the Advancement of Science, 1982.

"A Mirror on Their Lives: Capturing the Human Experience." In *Sharing Alaska's Oral History*, edited by William Schneider. Fairbanks: University of Alaska Press, 1983.

"Shooting a Buck." *Harper's*, January 1987.

"The Gifts." In *On Nature: Nature, Landscape, and Natural History*, edited by Daniel Halpern. Berkeley, CA: North Point Press, 1987.

The 1988 Western Wilderness Calendar. Salt Lake City: Dream Garden Press, 1988. (Excerpts from *Make Prayers to the Raven.*)

"Hunters and Animals in a Native Land: Ancient Ways for the New Century." *Orion*, Spring 1989.

"The Forest of Eyes." In *Alaska: Reflections on Land and Spirit*, edited by Robert Hedin and Gary Holthaus. Tucson: University of Arizona Press, 1989.

"When Civilization Ran Aground Aboard the Oil Tanker in Alaska." Opinion, *Los Angeles Times*, April 9, 1989.

"A Council of Trees." *Life*, May 1989. (From *The Island Within.*)

"The Island Within." *Parabola* 14, no. 3 (August 1989). (From *The Island Within.*)

"The Hidden Island." *Pacific Discovery*, October 1989. (From *The Island Within.*)

"A Mountain in My Hand." *Outside*, November 1989. (From *The Island Within*.)

"The Clearcut." *Harper's*, November 1989. (From *The Island Within*.)

"Gifts of Deer." *Alaska*, September 1990. (From *The Island Within*.)

"Alaska: A Glint in the Raven's Eye." *Wilderness*, Winter 1990.

"The Gifts." In *The Norton Book of Nature Writing*, edited by Robert Finch and John Elder. New York: W. W. Norton, 1990. (From *The Island Within*.)

"Tingiivik Tatqiq: The Moon When Birds Fly South (September)." In *Intercultural Journeys Through Reading and Writing*, edited by Marilyn Smith Layton. New York: Harper Collins, 1990. (From *Shadow of the Hunter*.)

"The Subsistence Cycle." In *Republic of Rivers: Three Centuries of Nature Writing from Alaska and the Yukon*, edited by John A. Murray. New York: Oxford University Press, 1990. (From *Make Prayers to the Raven*.)

"Ways of Knowing: An Interview with Richard Nelson," by George K. Russell. *Orion*, Autumn 1990.

"The Island Within." In *Blind Donkey* 12, no. 3/4 (1991).

"Exploring the Near at Hand: An Interview with Richard Nelson," by Rob Baker and Ellen Draper. *Parabola* XVI, no. 2 (Summer 1991).

"An Elder of the Tribe." In *Gary Snyder: Dimensions of a Life*, edited by Jon Halper. San Francisco: Sierra Club Books, 1991.

"The Forest of Eyes." In *Parabola* XVII, no. 1 (February 1992).

"The Forest of Eyes." In *Nature's New Voices*, edited by John Murray. Golden, CO: Fulcrum, 1992. (From *The Island Within*.)

"A Power Beyond Words." In *The Great Bear: Contemporary Writings on the Grizzly*, edited by John A. Murray. Bothell, WA: Alaska Northwest Books, 1992. (From *Make Prayers to the Raven*.)

"The Island's Child." In *Mountain Record* XI, no. 2 (1992). (From *The Island Within*.)

"Island Within," by Richard Leviton. *Yoga Journal*, January/February 1992.

"The Way of the Hunter: An Interview with Richard Nelson," by Jonathan White. *Sun*, May 1992.

"The Embrace of Names." *Northern Lights*, 1992.

"The Island Within." In *Left Bank*, vol. 2. Hillsboro, OR: Blue Heron, 1992.

"The Gifts of Deer." *Sun*, May 1992. (From *The Island Within.*)

"We All Live in a House With One Room." Introductory essay for John Muir's *Travels in Alaska*. New York: Viking Penguin, 1992.

"The Forest of Eyes." In *Beloved of the Sky: Essays and Photographs on Clearcutting*, edited by John Ellison. Seattle: Broken Moon Press, 1993.

"The Subarctic Wolf, 1983." In *Out Among the Wolves*, edited by John A. Murray. Seattle: Alaska Northwest Books, 1993. (From *Make Prayers to the Raven.*)

"Searching for the Lost Arrow: Physical and Spiritual Ecology in the Hunter's World." In *The Biophilia Hypothesis*, edited by Stephen R. Kellert and Edward O. Wilson. Washington, DC: Island Press, 1993.

"Understanding Eskimo Science." *Audubon*, September/October 1993.

"Bringing in the Storm." In *From the Island's Edge: A Sitka Reader*, edited by Carolyn Servid. Saint Paul, MN: Graywolf Press, 1995. (From *The Island Within.*)

"Moon of the Returning Sun." In *Last New Land: Stories of Alaska Past and Present*, edited by Wayne Mergler. Seattle: Alaska Northwest Books, 1996.

"Introduction: Finding Common Ground." *A Hunter's Heart: Honest Essays on Blood Sport*, edited by David Petersen. New York: Henry Holt, 1996.

"Forest Home: Taking a Stand for Conservation and Community." *Orion* 16, no. 3 (Summer 1997).

"The Gifts of Deer." In *The Portable Western Reader*, edited by William Kittredge. New York: Penguin Books, 1997. (From *The Island Within.*)

"The Hunt." *Utne Reader*, March/April 1998. (Adapted from *Heart and Blood.*)

"Discovering Alaska." In *The Harriman Alaska Expedition Retraced, A Century of Change, 1899-2001*, by Thomas Litwin, New Brunswick: Rutgers University Press, 2005.

"Hunting Wisdom: The Iñupiat and the Polar Bear." In *The Last Polar Bear*, by Steven Kazlowski. Seattle: Braided River, 2008.

"A River Morning: An Encounter with Bears, Salmon, and Ancient Trees" and "Listening to the Tongass." Audio chapter and natural sounds montage in *Salmon in the Trees: Life in Alaska's Tongass Rain Forest*, by Amy Gulick. Seattle: Braided River, 2010.

"Voices from the Land: Ancient Echoes on the Utukok," "The Old River," and "Surrounded by Caribou." Written and audio chapters in *On Arctic Ground: Tracking Time Through Alaska's National Petroleum Reserve*, by Debbie S. Miller. Seattle: Braided River, 2012.

NOTES

Solid Ground

"What wealth," Lopez asks: Barry Lopez, *The Rediscovery of North America* (New York: Vintage Books, 1992).

PART I: NIGLIK

Center of Gravity

I have been having considerable difficulty: Letter home, October 1964.

As his collection of critters grew: Flo and Bob Nelson saved all of Richard's report cards. Many of the cards have handwritten comments from teachers on the back.

The last grey light of day casts: A handwritten, undated copy of this journal entry was mixed in with other high school papers.

Stroke of a Paddle

I got the highest grade in the class: Letter home, July 20, 1960.

There was a day when the wind blew: Quoted from a typed, undated copy of "The Forgotten Prairie." A letter, dated January 25, 1962, from Cameron Wilson to Stewart Udall, says: "The Forgotten Prairie is a theme written for an English class by a friend of mine, Richard Nelson, who is a junior majoring in zoology at the University of Wisconsin."

Kusiq

The steam bark *Beluga*: Waldo Bodfish, *Kusiq: An Eskimo Life History from the Arctic Coast of Alaska* (Fairbanks: University of Alaska Press, 1991). This book is an oral biography of Waldo Bodfish, recorded, compiled, and edited by William Schneider in collaboration with Leona Kisautaq Okakok and James Mumiġana Nageak. The story of Kusiq's parents is told in great detail.

His entry on October 20, 1901: Captain Hartson Bodfish's logbooks from the *Beluga* are available in digital format from the New Bedford Whaling Museum (www.whalingmuseum.org/explore/library/logbooks/digitized-logbook-beluga-odhs_952A).

Kusiq was a friendly man: Wainwright Journal, September 9, 1964.

Richard shared his frustrations: Letter from Ken Taylor is quoted in entry in Wainwright Journal, October 4, 1964.

Cronkite and Caribou

I haven't heard from Rog: Recorded letter, October 6, 1964.

I made my own harnesses: Recorded letter, October 10, 1964.

Caribou clothing is the warmest: Recorded letter, December 6, 1964.

Richard's mind opened to the new sounds: Author interview with Ray Bane, October 29, 2016.

Today it was about zero all day: Letter home, October 26, 1964.

Last night they had a dedication: Recorded letter, November 22, 1964.

In late November: Recorded letter, November 22, 1964.

Borders

After two months in the Arctic: Wainwright Journal, September 4, 1964.

The Eskimo . . . dropped to one knee: Richard K. Nelson, *Hunters of the Northern Ice* (Chicago: University of Chicago Press, 1969), xxi–xxi.

The seal's head broke the surface: The introduction to *Hunters of the Northern Ice* opens with an Eskimo killing a seal, although as described here, Richard actually pulled the trigger, xxii.

Ice and Laughter

I have to admit: Letter home, December 20, 1964.
The accepted procedure is: Recorded letter, January 7, 1965.
Around here you feed your dogs: Recorded letter, January 1965.
This is another example: Recorded letter, December 1964.

Textured Tundra

The outstanding thing is the smell: Recorded letter, February 1965.
On the third breath: Richard K. Nelson, *Shadow of the Hunter* (Chicago: University of Chicago Press, 1980), 15–18.
It's getting warm: Letter home, March 31, 1965.
It's such a great big monstrous area: Recorded letter, November 1964.

PART II: MAKING PRAYERS

Kk'adonts'idnee

Catherine Attla, born in 1927: Richard K. Nelson, *Make Prayers to the Raven: A Koyukon View of the Northern Forest* (Chicago: University of Chicago Press, 1983), 19.
Catherine grew up in Cutoff: Catherine Attla interviewed by Mike Spindler in 1995. Audio interview archived at Project Jukebox at the University of Alaska Fairbanks, (www.jukebox.uaf.edu/mp3s/RavenStory/ca2.7.mp3).
In her early twenties: Huslia Journal, October 8, 1975.
Richard and Ray had no map: Author conversation with Ray Bane.
On this terrain the Athapaskan past: Nelson, *Make Prayers to the Raven*, 2.
On Richard's last evening in Huslia: This early interaction with Catherine was re-created from personal conversations with Richard.

Icy Cocoon

Worth more in real dollars than everything: Quoted in "The North Slope Oil Patch," by Neal Fried, *Alaska Economic Trends*, February 2018, (http://labor.alaska.gov/trends/feb18.pdf).

We're really loving the lifestyle: February 2, 1975, letter from Kathy Mautner to Flo and Bob Nelson.

In the following weeks, Ambler residents: Kobuk Journal, February 3, 1975.

Old Man Cleveland: Kobuk Journal, February 3, 1975.

I guess we're mighty lucky: Letter home, March 27, 1975.

When a man traveled: This passage was pulled from several different entries in the Kobuk Journal.

Forest of Eyes

On October 7—the year's first snow: Huslia Journal, October 14, 1975.

But being outside in this magnificent: Huslia Journal, October 27, 1975.

Luck

Most every page of Richard's Huslia Journal: Nelson, *Make Prayers to the Raven*, 26.

People who lose their luck: Nelson, *Make Prayers to the Raven*, 26.

In the absence of luck: Nelson, *Make Prayers to the Raven*, 26.

Chief Henry rubbed his whiskered cheek: Huslia Journal, October 13, 1975.

Skipping Heart

One night, dinner over, dishes done: Richard K. Nelson, *Make Prayers to the Raven* television program, produced by KUAC TV in 1987.

Catherine echoed these thoughts: Huslia Journal, November 2, 1975.

Catherine talked about how her grandfather: Huslia Journal, December 18, 1975.

Orphaned Ravens

I realized that no matter how hard: Author communication with Richard.

She went on to share a story: Huslia Journal, February 6, 1976.

What animal you want to know about: Conversation with Lavine Williams re-created from details in the Huslia Journal, January 1976.

Richard returned each afternoon: Conversation as remembered by Richard.

Chief Henry's Song

"Our ethos," wrote the historian Howard Zinn: Quoted in *Ethos*, Woody Harrelson (director), February 10, 2011.

I know my time is near: Huslia Journal, May 25, 1976.

Home

The place might not appeal: Letter home, August 31, 1976.

Got here yesterday: Letter home, April 19, 1977.

Just learned the sad news: Letter home April 19, 1977.

Dynamite

I'm not adapting well: Letter home, June 17, 1977.

Years later, Richard said: Richard K. Nelson, *The Island Within* (Berkeley, CA: North Point Press, 1989), 53.

The more I think about the future: Letter home, June 17, 1977.

PART III: ISLAND YEARS

Watch and Wonder

As I was living among: Nelson, *Make Prayers to the Raven*, 236.

We often remember ancient: Nelson, *Make Prayers to the Raven*, 246.

The fact that Westerners identify: Nelson, *Make Prayers to the Raven*, 246.
What is the raven?: Nelson, *Make Prayers to the Raven*, 248.

Blanket of Words

The deer must have lived: Nelson, *The Island Within*, 270–271.

Flood Tide

Thoreau wrote about Walden: Robert Finch, *The Outer Beach: A Thousand-Mile Walk on Cape Cod's Atlantic Shore* (New York: W. W. Norton, 2017), 59.
The nettling rain seems drawn: Nelson, *The Island Within*, 16–17.
I listen to the steady throb of surf: Nelson, *The Island Within*, 59.

PART IV: TRUE WEALTH

Keta

***Pilgrim at Tinker Creek*:** "Essential Books for the Well-Read Explorer," *Outside*, January 1, 2003.

Boots and Bike Locks

I stand to see the whole forest of stumps: Nelson, *The Island Within*, 54.
I hold few convictions so deeply: Nelson, *The Island Within*, 56–57.

Forgotten Language

I've now seen a wild deer born: Richard K. Nelson, *Heart and Blood: Living with Deer in America* (New York: Alfred A. Knopf, 1997), 352.
"Witness": W. S. Merwin, *The Rain in the Trees* (New York: Alfred A. Knopf, 1988).

Home, Land, Security

If this is true: Nelson, *The Island Within*, 17.

Covering the Sacred

This book is not only about the North Slope people: Nelson, *Shadow of the Hunter*, xiii.

We see a continuance in the present: Barry Lopez, *Rediscovery of North America*, 11.

Paths of Prayer

Richard envied his Koyukon teachers: Nelson, *The Island Within*, 25.

Richard went to the woods seeking solace: Nelson, *The Island Within*, 273–274.

True wealth—sanctity, companionship, wisdom, joy, serenity: Lopez, *Rediscovery of North America*, 21–23.

Remember you are not alone: Dalai Lama [Tenzin Gyatso] and Desmond Tutu. *The Book of Joy: Lasting Happiness in a Changing World*, with Douglas Abrams (New York: Avery, 2016), 273–274.

ABOUT THE AUTHOR

Photo by Liz McKenzie

As a life-long Alaskan, **Hank Lentfer** is familiar with much of the country Richard Nelson wrote about. Their friendship centered on a mutual love of wild country, writing, and listening to wild voices. Hank lives in the coastal community of Gustavus with his wife, Anya, daughter, Linnea, and extended family of dear friends. Winter, he writes and skis. Summer, he's either swinging a hammer, tending the garden, or recording birds. He is the author of *Faith of Cranes: Finding Hope and Family in Alaska* (Mountaineers Books, 2011).

recreation • lifestyle • conservation

MOUNTAINEERS BOOKS is a leading publisher of mountaineering literature and guides—including our flagship title, *Mountaineering: The Freedom of the Hills*—as well as adventure narratives, natural history, and general outdoor recreation. Through our two imprints, Skipstone and Braided River, we also publish titles on sustainability and conservation. We are committed to supporting the environmental and educational goals of our organization by providing expert information on human-powered adventure, sustainable practices at home and on the trail, and preservation of wilderness.

The Mountaineers, founded in 1906, is a 501(c)(3) nonprofit outdoor recreation and conservation organization whose mission is to enrich lives and communities by helping people "explore, conserve, learn about, and enjoy the lands and waters of the Pacific Northwest and beyond." One of the largest such organizations in the United States, it sponsors classes and year-round outdoor activities throughout the Pacific Northwest, including climbing, hiking, backcountry skiing, snowshoeing, camping, kayaking, sailing, and more. The Mountaineers also supports its mission through its publishing division, Mountaineers Books, and promotes environmental education and citizen engagement. For more information, visit The Mountaineers Program Center, 7700 Sand Point Way NE, Seattle, WA 98115-3996; phone 206-521-6001; www.mountaineers.org; or email info@mountaineers.org.

Our publications are made possible through the generosity of donors and through sales of 700 titles on outdoor recreation, sustainable lifestyle, and conservation. To donate, purchase books, or learn more, visit us online:

MOUNTAINEERS BOOKS
1001 SW Klickitat Way, Suite 201 • Seattle, WA 98134
800-553-4453 • mbooks@mountaineersbooks.org • www.mountaineersbooks.org

An independent nonprofit publisher since 1960

OTHER TITLES YOU MIGHT ENJOY

Faith of Cranes: Finding Hope and Family in Alaska
Hank Lentfer
"A love song to the beauty and worth of the lives we are able to lead in the world just as it is, troubled though it be."—David James Duncan

The Starship and the Canoe
Kenneth Brower
"In the tradition of Carl Sagan and John McPhee, a bracing cerebral voyage past intergalactic hoopla and backwoods retreats." —*Kirkus Reviews*

Swallowed by the Great Land and Other Dispatches from Alaska's Frontier
Seth Kantner
"Kanter's pull-no-punches, head-on stories are raw, beautiful, and unnerving." —*Orion*

Midnight Wilderness: Journeys in Alaska's Arctic National Wildlife Refuge
Debbie Miller
". . . describes vividly the wonders of this magnificent 19-million-acre preserve in Alaska's northeastern corner, from its coastal plain to its mountains, glaciers and rivers." —*Publishers Weekly*

Wild Shots: A Photographer's Life in Alaska
Tom Walker
"In his new memoir, Tom Walker turns the lens on himself for a rich romp through an adventurous life." —*Alaska*

www.mountaineersbooks.org

Stages of sled building — Kamotiḥluruk

two boards — 2" × 8" × 12'. Cut bow, cut length f
other end, usually 8 to 13 feet.

Drive 3 to 5 60 penny nails up into boards from below and/or
drill before driving nails. Gives strength to them — preve
splitting. Plane bottom + top surfaces with both boards fast
together.

cut out bow piece. Fasten with long nails, perhaps

Nail on cross slats — 1" × 4" boards, 24" to 40"
width of sled). About 20 such boards for 8 foot sled,

or longer sled. Nail in place, two nails per side.

il long strips of wood along the top of each runner
cross slats. Size of strips 1" × 2"

Nail strips of same size inside of runners + to cross
+ in middle

tach runner strips, hardwood for mid winter land travel or
spring + fall land travel and sea ice travel, to bottom of
— Nails and/or stove bolts to attach. Metal co
ed over hardwood strips + then removed in mid winter
rpose sled.

ach upstanders about 1' foreward of the rear of